PLAY ANY-THING

ALSO BY IAN BOGOST

PLAY ANYTHING

The Pleasure of Limits, the Uses of Boredom, and the Secret of Games

IAN BOGOST

BASIC
BOOKS
NEW YORK

BOOK DESIGN BY LINDA MARK

Library of Congress Cataloging-in-Publication Data

Names: Bogost, Ian, author.
Title: Play anything : the pleasure of limits, the uses of boredom, and the secret of
games / Ian Bogost.
Description: New York, NY : Basic Books, 2016.
Identifiers: LCCN 2016019144| ISBN 9780465051724 (hardback) |
ISBN 9780465096503 (e-book)
Subjects: LCSH: Creative ability. | Popular culture--Social aspects. | BISAC:
  SOCIAL SCIENCE / Popular Culture. | PHILOSOPHY / Ethics & Moral
  Philosophy. | PSYCHOLOGY / Creative Ability.
Classification: LCC BF408 .B566 2016 | DDC 306.4/8--dc23 LC record available at
https://lccn.loc.gov/2016019144

10 9 8 7 6 5 4 3 2 1

If you are immune to boredom,
there is literally nothing you cannot accomplish.

—DAVID FOSTER WALLACE, *THE PALE KING*

Contents

Life Is Not a Game

"LIFE IS A GAME." I'M SURE YOU'VE HEARD THIS LITTLE adage. As a philosopher who is also a game designer, I hear it a lot. It's the easiest way for people to make polite conversation when they find out I'm interested in the creation, use, and meaning of a contemporary technological medium like games, as well as big metaphysical questions like "what does it mean for something to exist?"

Like most aphorisms, it mostly feigns insight. "Life is a game," and so . . . what, exactly? It ends, eventually? It pits you in a challenge against others? Or it puts you in collaboration with them? Or even that you, as a proverbial player, can manipulate people and things as if they were pawns in a game? Maybe it means that life is fun like a game—unless it's not, of course, in which case maybe life is like a *bad* game.

Let me assure you that life isn't really a game, and I don't want to turn yours into one. Reality is alright as it is. It's just hard to see *what* it is. This book offers a perspective on how to live in a world far bigger than our bodies, minds, hopes, and dreams, and how to do it with pleasure and gratitude. I approach that topic through the lenses of game design and philosophy—and psychology, anthropology, science, art, design, entertainment, computing, and literature.

The lesson that games can teach us is simple. Games aren't appealing because they are fun, but because they are *limited*. Because they erect boundaries. Because we must accept their structures in order to play them. Soccer sees two teams of eleven players attempting to use their feet, torsos, and heads to put a ball into a goal. Tetris asks you to position falling arrangements of four orthogonally-connected squares in order to produce and remove horizontal lines. And yet the experiences games like soccer and Tetris create are far larger than those boundaries convey on their own. That bounty results from the deliberate, if absurd, pursuit of soccer and Tetris on their own terms, within the limitations they erect. The *limitations* make games fun.

What if we treated everything the way we treat soccer and Tetris—as valuable and virtuous for being exactly what they are, rather than for what would be convenient, or for what we wish they were instead, or for what we fear they are not? Walks and meadows, aunts and grandfathers, zoning board of appeals meetings and business trips. Everything. Our lives would be better, bigger, more meaningful, and less selfish.

That's what it means to play. To take something—anything—on its own terms, to treat it as if its existence were reasonable. The power of games lies not in their capacity to deliver rewards or enjoyment, but in the structured constraint of their design, which opens abundant possible spaces for play.

Play isn't unique to games—it's just easy to talk about play from the familiar vantage point of games. Play, generalized, is the operation of structures constrained by limitations. Maneuvering a soccer ball into a net without the use of hands and arms. Constructing patterns of lines using only the odd-shaped tetrominoes in Tetris. These constraints animate the games; they make them what they are. Play is not an alternative to work, nor a salve for misery. Play is a way of operating a constrained system in a gratifying way. This general act can apply to anything whatsoever—soccer and Tetris, sure, but also yard work and parenting, errands and marriage.

Over the course of this book, I will upset the deep and intuitive beliefs you hold about seemingly simple concepts like play and its supposed result, fun. It's not only that we don't know how to play effectively; it's also that our ordinary sense of the term is wrong. We think that in play we do what we want, that we release ourselves from external duty and obligation and finally yield to our clearest, innermost desires. We think we know what we want, and we believe that we are in control of our fates. But all of these beliefs are mistaken.

We are obsessed with freedom, but we are also miserable and bored, despite living in an era of enormous surplus. Instead of seeing freedom as an escape from the chains of limitation, we should interpret it as an opportunity to explore the implications of inherited or invented limitations.

We might even *need* to do this, lest we fall into the madness of refusing the world rather than embracing it. We have many worries, but most of all we are afraid. We're afraid of ourselves and our fates, sure, but worse: we are afraid of our world and its contents. When confronting something—a job or a love interest, sure, but even a ketchup jug or a Sunday afternoon—we worry that it might harm or disappoint us. We worry that it might fail to meet our expectations. We worry that it won't even stick around long enough that it's worth *having* expectations, and then we worry about having worried about it so much.

We need to slough off all those false fears that keep us from truly living, and to replace them with a new sense of gratitude at the improbable, delightful miracle that such a bounty of possible loves and ketchups affords us. Play is the missing tool we need to accomplish this feat, but it's not play as you know it. It's not selfish, thoughtless play, the play of "just playing around," but the deep, deliberate play of soccer and of Tetris. When we play, we engage fully and intensely with life and its contents. Play bores through boredom in order to reach the deep truth of ordinary things.

The ultimate lesson games give is not about gratification and reward, nor about media and technology, nor about art and design. It is a lesson about modesty, attention, and care. Play cultivates humility, for it requires us to treat things as they are rather than as we wish them to be. If we let it, play can be the secret to contentment. Not because it provides happiness or pleasure—although it certainly can—but because it helps us pursue a greater respect for the things, people, and situations around us.

PLAY ANY-
THING

ONE

Everywhere, Playgrounds

How to have fun in an age of boredom?
By meeting the world halfway.
More than halfway, even. Almost all the way.

Y EARS AGO, I WAS RUNNING AN ERRAND AT AN UPSCALE shopping mall in Atlanta, where I live. I was in a hurry, rushing from one store to another to meet up with my wife. The mall was crowded and bleak and I wanted to leave.

I had my young daughter in tow—she was four years old or so. She clutched my hand as I steered us quickly through the throngs of weekend shoppers. I was moving too fast for her small legs, and she was struggling to keep up.

But even as I felt her skipping between steps to keep up with me, I also felt her tugging me back intentionally, resisting my forward momentum, pulling me in another direction. When I looked down I saw why: she was staring straight at her shoes, timing her footfalls to ensure she stepped within the boundaries of the square, white tiles

1

lining the mall floor. The sensations I interpreted as pulls and tugs had been caused by shifts in her weight as she attempted to avoid transgressing the grout lines, while I pulled her forward and sideways around crowds.

Everyone will recognize my daughter's improvisation: it's a variant of "step on a crack, break your mother's back," a superstition of the late nineteenth century that developed into a children's game for sidewalks.[1] But my daughter's version adds intrigue and complexity: rather than resist or gripe about the admittedly unreasonable speed of my cadence, she'd chosen to subject herself to it. Since I was driving, so to speak, she didn't have to choose where she was going. This new freedom allowed her to focus on her feet rather than on human obstacles. But in so doing, she also surrendered control over her own forward motion.

Ordinarily, "step on a crack" makes no assumptions about its players' motion; you could trail behind a group, stand still while plotting your next step, or whatever. But for my daughter, my rapid and haphazard motion acted as a propulsion system. It was as if she were being conveyed through a carnival ride instead of by her own locomotion. She had to make quicker and more definitive decisions than she might have done otherwise. The result was pleasurable, vertiginous, challenging, and interesting.

She made up a game; she was "playing," we say, often dismissively. She made the most of a mundane situation. She turned misery into fun.

■■■

USUALLY, WE RELEGATE this sort of triumph to the ingenuity and resolve of children. Not yet worn down by the nuisances of life and not yet socialized into the stringent, boring norms of supposedly polite adult society, kids find joy until we beat it out of them through routine.

These days, psychologists and educators rejoin this tendency toward solemn efficiency. Can't adults find such joy too? Might they not even *need* to do so to thrive? But perhaps the unbridled shamelessness of scuttling akimbo between glazed porcelain tiles isn't meant to last past childhood. Instead, we can learn from the general practice rather than the specifics. Children aren't only less inhibited than adults; they are also less powerful, and smaller too. They may or may not be more open-minded and liberated than grown-ups, but they are forced to live in a world that wasn't designed for them, and one that is not primarily concerned with their desires and their welfare. And so children are constantly compromising, constantly adjusting to an environment that is clearly not theirs, not yet. That's wisdom, not innocence.

If only we could harness that wisdom and make use of it. Not only by dancing through malls, but also by approaching life with the attention and prowess my daughter exhibited that day—for her, anything could be coaxed into releasing meaning and pleasure and joy. I mean it, too: *anything whatsoever*, no matter how seemingly boring or stupid or meaningless—shopping and transit and work and chores and residential zoning disputes and tax preparation. After all, adults *also* don't live in a world designed for us. Climate, entropy, accident, crowds, happenstance, erosion, heartbreak—we are fools to think that we are in control of the universe. Children are right to allow the humility of their smallness to rule the day.

My daughter showed us the key: misery gives way to fun when you take an object, event, situation, or scenario that wasn't designed for you, that isn't invested in you, that isn't concerned in the slightest for your experience of it, and then *treat it as if it were*. As we'll learn in more detail, this is what play means. Play isn't doing what we want, but doing what we can with the materials we find along the way. And fun isn't the experience of pleasure, but the outcome of tinkering with a small part of the world in a surprising way.

Think about a musical instrument, like a guitar. When you *play* it, you don't do whatever you want—not at all. Rather, you do something very particular. You hold its fretted fingerboard in some patterns while strumming its strings in others. By manipulating the physical configuration of the device, you make it produce a subset of the infinite pattern of sounds we call music. And even if you don't know *how to play* the guitar, you can still *play with* it. Just holding a guitar as if you have mastered it is a kind of play, a make-believe that signals soul and nonchalance. Likewise, the hair-metal guitarist who swings his axe to trash the concert set also *plays* it in yet another way, wielding the solid-wood body as a demolition tool.

You can play anything, it turns out, like my daughter can play the ceramic floor tiles, like Hendrix can play the Fender Stratocaster. This book is about how to play, and why we need to. Play seems unserious and trivial, until playful encounters with familiar objects like floor tiles and guitars demonstrate that, no, play invites and even requires *greater* attention, generosity, respect, and investment than its supposedly more serious alternatives do.

Confronted with the arbitrariness of the world and all its contents, we are faced with a challenge: how to make do, how to find meaning, how to thrive and flourish even though the universe is ultimately indifferent. One answer is to resist, to abdicate, to reject our surroundings, possessions, and relationships as potentially insufficient and therefore untrustworthy. But another answer is to embrace the stupidity of mall floors and guitars and everything else—to allow the things we encounter to set the terms for our scrutiny rather than insisting that our own ideas and expectations should rule our experience.

OUT OF OUR HEADS

We still want to dance over the sidewalk cracks. But we've convinced ourselves that we can't, or we shouldn't. Not only because it's childish, but also because it's *boring*. We've done it all before: we've run our errands and filled our gas tanks and walked our dogs and mowed our lawns, over and over again. What a nuisance every day is, every day followed by the next. If only we could find relief for the suffocating boredom that we feel almost constantly.

In fact, boredom is a sure sign that we have tousled the hair of wonder and joy. Boredom sends up a flare: meaning exists here, boredom beckons, but *stranded* meaning. Meaning that requires rescue. Boredom is only insipid and lifeless if repetition must also be unimaginative, bland, and desolate. And yet repetition underpins so much of the joy in life: the delight in seeing the leaves turn again in autumn, along with the tender lamentation that summer is ending; enacting the rituals of winter holidays with children capable of doing something slightly new this time around; shoving off on an annual vacation, full of anticipation and vigor, and then returning full of exhaustion and prawns. The same weeks dissolve into the same weekends when we work on the same hobbies and partake in the same sporting events. Yet we assume that all of this supposedly *boring* repetition constitutes a full and rewarding life.

What we lack is a method for treating *anything* as inspiring like Christmas and vacation, autumn and weekend soccer. Boredom's affliction arises when we fail to know how to treat things with the admiration that leads to respect, and thereby to gratification. And not only to treat them that way in order to gain what we think we want, like the spoils of birthdays or the leisure of vacation. But also because treating something as it actually is—even a boring, insipid, commercial trip to the mall—releases secrets we might otherwise miss.

Heroism permeates ordinary life, in repetitions far smaller and weirder than the flow of the seasons and the years. In morning coffee and daily commutes, in grocery shopping and in yard work. Isn't it interesting that we're willing to call *fun* the hard, even miserable pleasure of playing a game like my daughter's on our shopping trip, but we refuse to call the hard, even miserable pleasure of the shopping trip itself *fun*? It's possible we fail to have fun not because of the world's ceaseless boredom and misery, but because we don't even know what fun is.

Boredom is the secret to releasing pleasure. Once something becomes so tedious that its purpose becomes secondary to its nature, then the real work can start. My daughter encountered that principle in the mall. There, smothered by the incapacity of being dragged on a stupid errand, she became so disgusted as to move beyond disgust, to find something—anything—that she could grasp onto as the bottomless pit of dark tedium cascaded over her. Otherwise, how would she even have noticed the grout and the tiles, let alone fashioned something new out of them?

She was having fun, but her fun emerged from misery. Fun isn't pleasure, it turns out. Fun is the feeling of finding something new in a familiar situation. Fun almost *demands* boredom: you need the sense that nothing good could possibly arise from an experience in order for the experience of finding something there to smolder with the hot pleasure of surprise.

Likewise, games aren't the opposite of work, but experiences that set aside the ordinary purposes of things. Things like felt tables and plastic cards, green meadows and round balls, sure. But also industrial ceramic floor tiles. Instead of asking the things what they can do for us, when we play we turn the tables on them. We get outside our own heads and accept the objects, circumstances, and people that surround us. Mindfulness is the practice of accepting our own thoughts and feelings, but what good is it if we accept only ourselves? We need a means to accept *other* things. A *worldfulness* to complement—or even to replace—the trend of mindfulness.

Why? Because our imperfect minds have led us astray. And not only because we think we can find gratification by listening to them more closely, but also because we trust that satisfying our minds and bodies should be our primary focus. We become impatient and irate when the world we encounter doesn't correspond with our own hopes and wishes. We start to regard everything as an obstacle, nuisance, and even a threat. What, after all, could ceramic tiles ever do for our souls, eternal or otherwise? Better, it would seem, to reject or dismiss or ignore them so that we can get on with better things. We become so busy foregoing or discarding things that we wouldn't know what "better things" are anymore. We fear what we do not have, and so we purchase and acquire more and more. But then we fear what we possess, either because it proves less satisfactory than we expected or because we feel guilty for having acquired it. And so we purge our surroundings in a futile attempt to replace consumerism with asceticism.

...

I FEEL BORED and irate and impatient as much as anyone. Like my daughter, like you, like everyone. But I'm not sure I manage my own trials with the adeptness of my daughter in the shopping mall. Play is harder than it seems. It doesn't happen automatically. If you compare my daughter's tolerance for nuisance and displeasure to my own, her reaction exceeds the proficiency with which I address, theorize, manage, and resolve ordinary ordeals. I suspect the same would be true for you, too. Instead of taking things in stride, instead of transforming them from insufferable to agreeable, our default approach tends toward frustration, overwhelm, anger, and disgust. Rather than accepting the invitation to play, we reject the call as insufficiently compatible with our predetermined needs and wishes.

The worst part is that we kind of like this selfish response. We even revel in it. Suffering and torment may raise our blood pressure and eventually shorten our lives, but they are also marks of social maturity. They signal sophistication. In a famous Kenyon College commencement speech that was later published as the short book *This Is Water*, the late American writer David Foster Wallace attempts to impart this lesson on twenty-two-year-olds. "There happen to be whole large parts of adult American life that nobody talks about in commencement speeches," says Wallace. "One such part involves boredom, routine, and petty frustration. The parents and older folks here will know all too well what I'm talking about."[2] He proposes a scenario to illustrate the matter. You're tired and stressed after a long day's work, but on the way home you remember that you need to stop for groceries. You get stuck in traffic. Then the supermarket is crowded, badly lit, and filled with Muzak. By the time you finally check out and awkwardly fit all your bags of groceries into the car, you're furious and frustrated and defeated. Then you repeat these kinds of frustrating episodes most of the day, every day, for most of your life. *Gulp.*

Wallace suggests that we have to resist seeing ourselves as the central, oppressed actors in a ghastly drama. The answer, for Wallace anyway, is to decenter yourself, to resist this selfishness as a default mode of thinking. Rather than seething at the slow woman with the coupons waiting for the check approval, you can instead forbear and desist. "I can choose to force myself to consider the likelihood," Wallace writes, "that everyone else in the supermarket's checkout line is just as bored and frustrated as I am, and that some of these people probably have much harder, more tedious or painful lives than I do, overall."[3]

It sure sounds good, doesn't it? But, alas, Wallace's alternative to the madness of default selfishness is an equally soul-destroying, utterly boundless hypothetical empathy. He advises us to retreat *further* into the self, which makes it more difficult truly to accept the woman at the checkout—or anyone or anything else, for that matter.

Imagine striving to consider all the possible scenarios and contingencies among everyone and everything that surrounds you: the pallid woman in the supermarket line who might have spent the last year battling breast cancer, or the giant, aggressively-driven, gas-guzzling SUV that might transport a decent working-class father who was recently laid off and is struggling to resist a descent into depression after losing his identity as breadwinner. Spinning fictional explanations for every impulse and sensation might help reduce our sense of entitlement and centrality, but only if we can escape from the prison in our heads. No quantity of mindfulness and secular spiritualism will be sufficient if, at the end of the day, we also require that all meaning originate from within our own addled, overtaxed brains. Wallace's standard—assuming that everyone has "harder, more tedious or painful lives"—goes further still: beyond inventing meaning, our burdened skulls apparently must invoke the most drastic situation in order to subordinate our private feelings to the circumstances we encounter. A rat race for worst-case scenarios.

It's insane to think that we'd have to *make up* fake stories when the world is so replete with real stuff waiting for us to notice it— stuff like rectilinear shopping-mall floor tiles, Gibson Les Paul Studio guitars, the knobby stem-necks of tangelos, cans of Pringles machine-formed potato chips, the formal constraints of a tweet or a sonnet, and countless other ordinary things that this book explores. To treat things with respect and intrigue, we don't need to understand their motivations and inner lives—whatever knowing the inner life of a tangelo or a floor tile would mean.[4] We just need to pay enough attention to discover what they do and how they work—to discover what they obviously and truly *are*—and then to make use of them in gratifyingly novel ways.

And yet *missing the obvious* is precisely the point of Wallace's Kenyon address. The title *This Is Water* comes from an introductory parable in which an older fish exchanges a pleasantry with two young ones, asking, "How's the water?" The older fish swims on, and one of the younger fish responds to his compatriot, "What the hell is water?" The point of this story, writes Wallace, "is that the most obvious, ubiquitous, important realities are often the ones that are the hardest to see and talk about."[5]

COMMITMENT, ATTENTION, AND CARE

It's ironic that the young fish in Wallace's parable don't know what water is despite their lives' reliance on it. But even worse than not knowing what surrounds us every day is to know and *still* to mistrust it. To embrace the situational irony of the fish as *intentional*. "Psh, whatever it is, I certainly don't *need* water," the contemporary fish scoffs. "I'm just using it until something better comes along."

Irony keeps reality at a distance. It has become our primary method for combatting the external world's incompatibility with our own desires. Today's irony uses increasingly desperate efforts to hold everything in between welcome embrace and sneering mockery. Irony is the great affliction of our age, worthy of its own disorder.

I call it *ironoia*, an idea I'll explain in full soon enough. Paranoiacs fear that *people* are out to get them but ironoiacs fear *anything whatsoever* has it in for them. It's a condition brought about by the sense that anything could go wrong at any moment, and therefore everything is duplicitous, untrustworthy.

It's no wonder we feel mistrust: think of all the obvious reasons we have to be fearful. September 11 and the amorphous war on terror that follows it without end; the global economic collapse of 2008 and its subsequent era of precarity, austerity, and inequality; the ever-accelerating drive toward consumerism despite that precarity, and the associated guilty or ascetic backlash; the boundless sea of entertainment on television and Netflix and social media and apps and websites, all scrabbling for a tiny crumb of our attention. The conditions are ripe for fear.

But we have allowed that fear to overtake all other drives. And when pushed to its limits, fear turns into selfishness. Fear assumes that we ourselves—no matter who we are—deserve to be the central actors in the universe's drama. Irony is the risk management strategy that accompanies selfishness, whether in commercial form as materialism or in spiritual form as mindfulness. By holding everything at a distance, we trap ourselves within our imperfect minds.

Irony doesn't protect us; it only makes things worse. The world is too big and too weird to hedge against with a psychic insurance policy. For every trinket or foodstuff or invitation or Internet post one successfully resists, a hundred, a thousand, a million others are waiting in line to take its place. If not shopping then traffic. If not traffic then work e-mails. If not e-mails then plumbing repair or parenting or taxes or bagels. To pretend that the world only exists in one's head is a madness condemned to reproduce itself forever. The error mistakes the big, weird world outside our heads for a world built to be housed inside that head, in our comparatively tiny minds.

...

THERE ARE ALTERNATIVES. They follow in the footsteps of my daughter's playful game, but they can be conducted anywhere, upon anything, and by anyone. First, pay close, foolish, even absurd attention to things. Then allow their structure, form, and nature to set the limits for the experiences you derive from them. By refusing to ask what could be different, and instead allowing what is present to guide us, we create a new space. A magic circle, a circumscribed, imaginary playground in which the limitations of the things we encounter—of anything we encounter—can produce meaningful experiences.

Let's be clear: I am not suggesting that life is a game or that work should be replaced by play. When faced with a strange, scary, and contemporary technological medium like games, people want either to vanquish it as a threat or to bottle its black magic for use elsewhere—in homes or schools or workplaces. But games aren't magic, and the most special thing about them isn't unique to them anyway—their artificial, deliberately limited structures teach us how to appreciate *everything else* that has a specific, limited structure. Which is just to say, *anything whatsoever.* Play isn't our goal, but a tool to discover and appreciate the structures of all the malls and fishbowls we encounter.

Once you look for examples of things to play with, you'll discover that nothing is impervious to manipulation. Sometimes eking out clues about how to use a thing involves a good measure of boredom, however. Counterintuitively, the more familiar and less immediately engrossing something becomes, the more it is subject to play. You'll soon become startled noticing how frequently you manipulate the material world within its natural limits. Stuck in traffic, your fingers dance across the radio presets. Held hostage on a conference call or webinar, your hand idly doodles variations of the same rhomboid shape upon the backside of the agenda. You load the dishwasher by the special method you have devised to fit all the short glasses under and atop the folding shelf so as to free up more space for larger glasses and bowls. And then you hide your deployment of said strategy from your spouse, who has no patience for overstuffing the appliance and thereby causing it to clean less effectively.

The next step: to take hold of these tiny moments and to expand them into long-term commitments. Take coffee. You forego the one-button pod-style coffee brewer for a manual or semiautomatic machine steeped in tradition and history rather than waste and commerce. You freshly grind your espresso beans. You fluff and tamp them to a measure and density discovered over many other mornings. You time the flow of temperature-regulated hot water through your machine's group head to produce a twenty-seven-second pull that balances sourness against bitterness in this particular roast, one you settled on after experimenting with several in small batches.

New equipment and techniques add to your sense of investment, mastery, and understanding of the seemingly ordinary practice of brewing coffee. A kitchen scale might allow a more precise measurement of grounds and thereby produce a more reliable extraction time. Perhaps you invest in a bottomless portafilter that allows you to monitor the surface tension on the stream of coffee. This monitoring facilitates revisions in your method for more even extraction. Thanks to the Internet, you discover that such an approach is called the Weiss Distribution Technique (WDT), a comically official-sounding name for a practice that amounts to breaking up clumps of grounds for a more uniform distribution in the filter basket.

Then, over time, as you acclimate to the process and the tools, the fun comes from removing them again. From eyeballing grind measures by feel, from experience. From timing your draws by intuition rather than by science. Later, as the habit of expertise grows too familiar, you try a new coffee blend that requires a new grind, a new tamp, a new temperature, and a new extraction length to produce a newly tunable taste.

This example might seem foolish or even pompous. We've trained ourselves to see commitments as affectations, and only to pursue a commitment ironically so that we can cast it aside if fear overtakes us. But foolishness signals that you're on the right track. Fun comes from the attention and care you bring to something that imposes arbitrary, often boring, even cruel limitations on what you— or anyone—can do with them. Worldly limitations impose a new and welcome humility, for they force us to treat things *as they are* rather than *as we wish them to be.*

Given the tyranny of boredom, fear, consumerism, asceticism, irony, and all our other failed strategies for making the world tolerable, we require such a shift. The great tragedy of Wallace's life—a lifelong sufferer of depression, he committed suicide at age forty-six—isn't only that he killed himself: it is also that he was unable to invent a tolerable, lasting mode of living during the years he eked out of the universe, a mode of living that truly allowed the selfish mind to live amidst the great outdoors. He was right about the fish: they need to learn to see the water. But then they need to do something with it.

THIS IS WATERING

Speaking of the outdoors: to be honest, I'm having trouble concentrating as I write today. I'm thinking about my lawn. Five weeks ago, I underwent an extensive renovation of the front yard: releveling the terrain; replacing the grass; and relocating the existing, poorly located plants, shrubs, and young trees, and adding several new ones.

The grass was the big project. Fungus that probably spread when I failed to clear a big snow two winters ago killed most of it, and poor runoff from a difficult, suboptimal grade from the street cultured the fungus. So we installed new sod, and I've spent the last month working to establish and culture it. If you've been through this process yourself, you know that it's a little like having a newborn baby or puppy: it requires eagle-eyed care and regular attention, mostly to ensure the new lawn doesn't dry out, and then that it doesn't drown, and then finally that it can find irrigation on its own from the depths of the soil, absent the hose or the thunderstorm. Like soccer or Tetris, lawns have fixed structures that amateur landscapers like me cannot ignore.

Four weeks in, I exchange text messages with my landscape designer, Mark, for some input. I was soon planning a trip to the nursery to pick up a successor to a *Euphorbia mellifera* (some folks call it honey spurge, a little pom-pom of an evergreen shrub) that had succumbed to a brutal slug attack.

Mark offers me some unbidden advice about weaning the new lawn back to less frequent waterings. He suggests reapplying preemergent weed control and, in the process, fertilizing to increase the density and verdancy of the zoysia (the variety of grass we'd installed). Mark recommends Scotts Turf Builder with Halts, which is harder to find than seems reasonable, given it's a fairly ordinary weed-and-feed product. Its rarity probably stems from the fact that the turf builder is more commonly applied in spring rather than summer. However, zoysias are warm-weather grasses that can only be cultured anew in the hottest months. I can't find the stuff locally so I finally order a bag from Amazon (you can get anything at Amazon; it's the new Walmart), along with a jar of Bug-Geta Snail and Slug Killer so I don't lose another euphorbia to the unsettling mollusk incursion. Two days later, you would have seen me fervently fertilizing out front, eager to finish the job before leaving town for a short business trip.

Upon my return, I spy a yellow patch just beyond the shrubs. All the blood sinks into my feet and—it feels this way at least—seeps from under my toenails and into the tufted soil beneath them. Huge swaths of yellow now crisscross the once-virescent fleece of the lawn. Like scars from some sort of torture, I think, before realizing that *I* am the torturer, and the scars the result of my apparently cruel and inept application of Scotts Turf Builder with Halts.

Chemical fertilizers, you see, contain salts that leech water from the soil. Overapplication effectively strangles the roots, resulting in yellowed, strained shoots—or worse, brown and dead ones. I briefly entertain casting blame onto Mark or the ambiguous Scotts fertilizer instructions before remembering that I had looked for the small fertilizer spreader I'd bought in a previous season, only resorting to hand-casting after I couldn't find it and, pressed by the imminent out-of-town trip, had decided to get the job done right then. It's not so much that I only have myself to blame—rather, blaming myself is a stupid way of responding to a situation that now has nothing to do with me or Mark or Scotts, and everything to do with the moisture, mineral, chemical, and especially salt content of my lawn.

And therein lie answers, and contrasts. I could spin stories about the zoysiagrass the way Wallace does for the supermarket-goers, and such an idea might make good metaphysical poetry. But more productive—for me and the grass alike—would be to understand and respond to the material situation it presents, and to attempt to interweave that reality into my own emotional state (one of admitted horror and, relatively speaking, true and unironic despair). To work with the grass on its terms rather than on mine.

In this case, it means flooding the lawn with as much moisture as it will take in an attempt to dilute the chemical and salt content of the soil. Then, trimming down and removing the yellow-browned blades in order to free the soil for new stems to spread rhizomatically—by spreading out their shoots horizontally. Then watering the edges of the affected areas to encourage new growth, but also persisting with the irrigation-weaning plan to strengthen the roots. And then to wait for the bare patches to fill. Zoysia grows slowly but steadily. A lawn's time horizon is not the same as a man's, and it does not become impatient like I do as it lugubriously spreads new runners from carpet to soil and evens the green of my lawn.

I must find a way to accept the disaster. Frustration is one way of interpreting the difference between what I wanted and what lawns do. Another way is to acknowledge that the world is outside my head rather than within it.

And, to be frank, in this effort I still don't manage to match the prowess of a four-year-old skipping on ceramic tiles. I internalize all the guilt and idiocy I just convinced you I'd dodged, swelling up with anger and embarrassment. I storm and grumble about it. My wife and I don't talk about my foul mood, and we won't until she reads this very paragraph. I'm still angry at myself and Mark and Scotts and the cruel world that's clearly pointing all its wrath straight at me, like some intercontinental missile whose warhead is made of spleen rather than plutonium. As I'm out watering the yellow spots, on my knees like a drunk who's lost his keys, my only remaining hope is that Joe and Eileen from down the block don't pass on their usual morning walk. Joe had seen me out there just the other day and congratulated me on my great new lawn, before he and Eileen pressed on fast like dignified show horses.

But of course, just then, Joe and Eileen do walk by. Of course. I'm bent over, back to the street, hose unfurled and disgorging. It must look like I'm trying to cover up a murder scene.

"Ian?"

I look up. I'm still hoping he doesn't notice.

"Are those fungus patches? Because, you know, there's antifungal—" I cut him off. I know all about fungus; it's why I had to replace the lawn in the first place.

"It's . . . " My brain shuts off searching for the word. All my neurons are smothered by shame, by the stupidity of being ashamed of a goddamned lawn. "It's fertilizer burn," I manage.

"Oh," Joe nods, scrunching his lips to signal either earnest empathy or secret schadenfreude, and off the two go, their thick, white manes proud and upright. I squeeze the hose's trigger and watch the nozzle's jet penetrate the already-sopping mat of flaxen blades at my feet.

"This is water," I say aloud.

...

WHAT'S THE DIFFERENCE between my daughter's playful encounter with the mall and my sod-and-fertilizer mishap? Or better, what would I need to do in order to experience my temporary lawn tragedy as the wondrous miracle that my daughter found in the mall floor rather than the shameful nuisance that I felt standing in my yard before Joe and God and the universe?

You can find the answer by inverting David Foster Wallace's hypothetical solution to trials on the road and in the supermarket. Getting the people and circumstances of daily life out of his head (and ours), and returning them to the world.

Wallace's answer to the day-in, day-out monotony of ordinary life involves ceding the whims of the ego in recognition of the collective. Through sacrifice. "Real freedom," he writes, comes from "being able truly to care about other people and to sacrifice for them, over and over, in myriad petty little unsexy ways, every day."[6] But remember that, for Wallace, this care and these sacrifices don't derive from the needs and wants and circumstances of all those other people. Rather, they arise from the newly burdened imagination of the self. Helping the cancer-surviving coupon-clipper in the checkout line is possible if you know she's a cancer-surviving coupon-clipper rather than an acrimonious sourpuss, or even an ordinary, humdrum nobody. But if you've invented that reality in your head, then your fingers will slip right through it. It's like playing a guitar made of water instead of wood.

The gravest mistake we make about play is thinking that it is unbounded, evanescent. Whether as amusement or as diversion, play becomes a desirable alternative to torment. Not work but play. Not structure but play. Not limitation but play. But play is always structured; you always play *with* something. The soccer athlete plays with a combination of physical prowess, the physics of the ball, and the tactics of the opponents. The Tetris player plays with the position and orientation of tetrominoes to make them interlock, perhaps while waiting in line at Wallace's supermarket. The yard caregiver (that's me) plays with the connections between plants and climate and chemicals and machines.

Acknowledging, understanding, and accepting structures of all kinds—that's the difference between my daughter's joy in the mall and my irritation on the lawn.

PLAYGROUNDS, WHERE PLAY TAKES PLACE

Perhaps an old Zen master fish could look with great attention and *just see* the water. But the rest of us need help. Games are especially good at showing us what they are made of. A ball and two goals, no hands. Four squares stuck together, falling over time. But it's not only games; everything has borders and contents, edges and materials. Art, literature, software, floors, lawns, freeways, supermarkets. Everything.

Structure. That's the trick. Learning something about the proverbial water in which we're swimming and responding to it—both tasks demand a structure we can see and with which we can engage. Yes, true, acknowledging that we're immersed in things without noticing them is the first step. But that's the easy part. The hard part is working with things, really digging into and making use of them— or even just letting them be, but recognizing something new in the suffocating, familiar depths to which you and others already have put them to use. We nickname this experience "fun," and fun turns out to be fun even if it doesn't involve much (or any) enjoyment. Deliberateness and novelty are enough.

My daughter's mall game illustrates these principles. She saw and acknowledged the tiles, which are separately laid and grouted for the ease of manufacture, transport, installation, and maintenance. But rather than allowing that material distinction to recede into the background, to become mere substrate for our far more urgent pursuit of retail commerce, she made the tile/grout pairing the focus of her attention. She added to them the speed of her gait as pulled along by me, my hand and body as it attached to and pulled her to and fro erratically, the shape and size of her feet, the traction or slipperiness of her shoes, the vectors along which other mall-goers traveled in relation to her future path, and so forth. She had to *increase* her attention to detail in order to play, which runs counter to our ordinary conception of play as a release of attention and responsibility.

Then, even though I was merely an accessory to her game rather than a party to it, she forced me to recognize and acknowledge the space she'd created. The tiles, the grout, her shoes, and so on—I became newly aware of these things simply by virtue of attending to her indirectly. We must seek to capture that magic everywhere, in everything. Not the pleasure of realizing our own goals—as if we even know what they are or ought to be—but the gratification of meeting the world more than halfway, almost *all* the way, and reaping the spoils of our new discoveries made under the sail of generosity rather than selfishness.

■ ■ ■

HOW? WE CAN start with an easier question: where does one play? On a playground.

In its customary sense, *playground* refers to a recreational area, usually outdoors, expressly defined for children's play. But in a metaphorical sense, *playground* describes the place where play takes place, no matter the type of play. An Appalachian hiker might call the Blue Ridge Mountains her "playground," or an haute cuisine chef might call the cramped quarters of a restaurant kitchen his "playground."

Every playground has two basic properties, which are two sides of the same coin: boundaries and contents. Playgrounds are bounded in order to delineate the space where the playful behavior takes place and where it stops or ends. In the chef's kitchen or the park's play area, upon the chessboard or atop the Tetris playfield, those boundaries are clear and literal. The kitchen has walls, of course, and maybe even swinging doors that separate the preparation area from the dining area. The play area might be bounded by the cartography of a city block in which a park is located, or it might further be enclosed by a curb or a border that divides a sandboxed jungle gym from a larger green space.

The Blue Ridge Mountains, by contrast, are an amorphous region of the east-central United States, extending from northern Georgia into Pennsylvania, over a distance of hundreds of miles. Technically, the term refers to the eastern range of the southern Appalachians, but our hypothetical hiker need not draw us a map in order to describe the general scope and range of the area she intends to delineate, with its characteristic low-slung, bluish hue. The loosey-goosey nature of this distinction is intentional; geographers call them *physiographic* regions, which are defined by their geological structure and historical development.[7] Those features often establish the squishier characteristics that we tend to associate with physical places that aren't defined by rigid political or physical boundaries. A conceptual rather than a physical border.

Once a boundary is drawn, real or conceptual, a playground's contents are set into relief against all the materials the boundary excludes. The park's play area contains the sand, swing set, slide, and merry-go-round. The physical boundaries of the kitchen contain the line, pass, staff, stations, equipment, and foodstuffs. And the Blue Ridge Mountains contain the towns, roads, trails, scenery, wildlife, flora, climate, and outdoorsman activities that are at the disposal of our imagined Appalachian hiker, whose status as hiker helps us understand the boundary she means to erect by naming this particular physiographic region.

My daughter's "step on a crack" game demonstrates how flexible and arbitrary a playground's boundaries and contents can be. It's not the mall that is her playground, nor some hypothetical meter-wide circle surrounding her body and mine. A playground can be a hybrid of physical and conceptual worlds. Imagine drawing a magic circle that starts at her center and then radiates out into the material world, lassoing and enclosing some of its contents while excluding others. The tile, the grout, her feet, my hand, our motion, the foreign bodies swirling—all these things became party to her game by virtue of her containing them within a flexible boundary. A playground is a place where play takes place, and play is a practice of manipulating the things you happen to find in a playground.

Such is the difference between play and ordinary life. Freedom or even enjoyment doesn't distinguish the one from the other. Rather, play comes from the deliberate operation of the things among which we find ourselves. And those things are constantly shifting and re-configuring themselves, like cars in traffic, like leaves scattered in the autumn winds, or like frenzied shoppers and consumer goods in a grocery store or a shopping mall.

Wallace's answer is to leave things to their shiftiness, but then to rationalize their behaviors through secret explanations. This is the mania of selfless abdication: I can never fully satisfy the world, but I can think about doing so eternally. Another, more common answer is to try to pin everything down, to wait for your career or your lawn or your afternoon or your marriage or your shopping trip to settle into comfortable stasis before communing with them. This is the mania of selfish irony: the world can never fully satisfy me, so I will hold it at arm's length forever.

Wouldn't it be easier and more productive to work with the objects, people, and situations we encounter? To use, understand, and appreciate them for what they are rather than for how they make us feel about ourselves?

...

WHICH LEADS US back to my lawn. How might I follow my daughter's lead and find comfort in a thing ultimately out of my control, but over which I can still exert some limited control? By building playgrounds within it. Playgrounds, plural, because once you look, you'll see endless ways to circumscribe and address the perverse and fundamentally idiotic burden of owning and managing a meadow you have coerced to flank your property.

We don't see them all at once. We stumble into some by accident, and then, over time, new ones reveal themselves.

I start by buying, borrowing, and operating the equipment necessary for basic lawn care. The bread-and-butter stuff, like the mower and the edger and the rakes and the lawn bags and the gloves and the sweatband. I learn the ins and outs of these tools and develop an expertise for them, perhaps even at the risk of consumerist accumulation.

But I can't manage all the playgrounds, so I outsource to someone like Mark, whose skill and experience far exceed my own. It costs me some money, which means that I also must budget for a lawn instead of a band saw or a more capable espresso machine. In exchange I get time and peace of mind, but I pay a price for the privilege. In delegating, I miss out on learning something new about lawns in general and my lawn in particular. I don't have seasonal equipment, like the rolling spreader or the weed-and-feed or the winterizer, nor the expertise to use them. This lack of tools and skills leads me to experience the frustrating but equally real and—retrospectively, at least—fun experience of having encountered and come to terms with fertilizer burn.

Playgrounds overlap, both containing and excluding the contents of other, neighboring ones. Part of discovering something new about a thing like an industrial floor or a lawn is knowing what choices one has in relation to them.

One of the choices doesn't involve the lawn much, but only my own wallowing in shame and embarrassment, as my encounter with Joe exemplifies. Forget the grass, *I'm* the one who's been injured here. Look how embarrassing these stripes of yellow are in my otherwise perfect American lawn!

For a moment, I externalize all that angst. It's Mark's fault, or the Scotts Company's! Why didn't Mark tell me exactly what I needed to do in intricate detail? Why were the Scotts Turf Builder with Halts instructions unclear or incomplete? Someone is to blame, and it's not me. "Where's Mark's contract?" or "I'm going to write a letter!" dads everywhere might intone. (Don't worry, I didn't write a letter.)

Suddenly, the ascetic arguments against home ownership and the ecological ones against landscaping feel newly relevant. Why do I even have a lawn? What a waste of water and space and money, and at a time when so many have so little, and when climate-change-driven drought and superstorms promise to forever unseat the silly, midcentury dream of home ownership and lawn-care obsession. I could sell the property, rent a more modest one within biking distance of work, and donate the time I would have spent doing lawn care to local cycling-ordinance advocacy.

Eventually, I settle on a different playground, at least for the moment. Its boundary draws when I accept my error and redirect my focus toward the lawn rather than my feelings about it. Its contents: the physical, chemical, and environmental properties of zoysiagrass, salt, water, and soil; the process of determining how to triage and treat the condition; and commitment of the time and effort required to carry out the remedy. It is, I hope, a more humble choice. One that abdicates the need to see myself as victor over a calamity I had created (or avoided). Instead, I commune with the lawn to understand something new and subtle about it, and to carry that experience into my and its future.

But the strange truth is that *all* of these playgrounds offer promise, meaning, and even fun. Even the supposedly gloomy playgrounds, like pondering some preposterous litigation and wallowing in my own gloom, were worth encountering. After all, accountability, consumer protection, the legal system, economics, humiliation, neighborhood rivalry, and the inner demons of one's shadow personality are all materials no more or less worthy of understanding and manipulating than are soccer pitches or lawnmowers or bags of Scotts Turf Builder With Halts.

Perhaps the problem comes in thinking that there should be one answer. Mistaking playgrounds as fixed slots for our selfish personas, as if everything that happens in the universe is just some potential input for a Myers-Briggs personality assessment or output from a BuzzFeed quiz.

Playgrounds are not thrones built for our proud gratification, but configurations of materials. They are not in our heads, but in the world. The first step in enjoying them is to stop worrying about our possible roles within them, and instead to allow lawns and malls and soccer pitches to show us their desires.

■■■

OUR WORLD IS jam-packed full of splendor and mystery, most of which we never notice as we ply the demands and dissatisfactions of our selfish lives. And even when we find mechanisms for relief—Buddhist mindfulness or libertarian objectivism, sermonic asceticism or unbridled consumerism—they turn our attention inward rather than outward. They tell stories about the bodies and minds we wish we occupied rather than offering us tactics to live amidst the world as it really is. Playgrounds aren't things we create so much as structures we discover. They are peculiar configurations of materials that otherwise go unnoticed, unseen, unloved, and un-appreciated. They're scattered everywhere, stacked, overlapping, ex-erting their machinations without us, but available for our address and manipulation, if we draw a magic circle around their parts and render them real.

The good news is that playgrounds' pervasiveness makes them incredibly easy to find once you start looking. And once you do, you'll see them everywhere. And once you see them, you can prac-tice using them, ratcheting up the skill with which you identify and manipulate all the other playgrounds you'll discover subsequently.

Living playfully isn't about you, it turns out. It's about everything else, and what you manage to do with it.

TWO

Ironoia, the Mistrust of Things

If paranoia is the fear of people, ironoia is the fear of things.
Today, we fear everything, so we keep it all at a distance.
But we always fail, because the world is too abundant to hold at bay.

YOU CAN'T FIND YOUR KEYS. THEY'RE NOT ON THE HOOK, not on the desk, not on the nightstand. You're already running late, and warm panic has begun to radiate from your middle out through your ears.

The German sociologist Niklas Luhmann once observed that the simple act of asking yourself, "Where did I put my keys?" performs unexpected magic: it transforms the world into a catalog of possible key locations.[1] Under the couch, somewhere the dog or the baby moved it, in the grocery store parking lot, under the seat on the train, in your very hand. It doesn't seem like it, but losing your keys fashions a playground as much as skipping across shopping-mall tiles does. It draws a boundary, separating your house right now from your house a moment ago, before you knew you had lost your keys. Suddenly, rooms become litanies of hiding places for keys rather than places to cook, eat, sleep, or watch television.

Once you finally find the keys, the boundary retracts, but the playground doesn't disappear, not entirely. It sticks with you. You make note of the places you thought you might have left the keys and store them away for later reference or present caution. You'll lose your keys again, after all. In the meantime, you communed with all those spots you otherwise overlook—the folds of the jeans in the hamper, the accumulation at the bottom of your purse, the neglected space between the car door sill and the seat where your hand doesn't quite fit.

We don't normally think of the panicked inconvenience of losing our keys as a playful experience. It's nerve-wracking. Miserable, really. But if mall floors and lawns and morning coffee routines can be playful once we choose to acknowledge, respect, and work within their boundaries, then anything else can be playful too. Play is everywhere, waiting patiently for us to bother to notice—let alone to partake.

I resolved to test the matter. Where might the greatest diversity of playgrounds meet the most limited inclination to see and engage them as such? The mall wasn't a bad start, but people did (and still do) choose to go there, to shop or walk or eat or people-watch. Wallace's loathsome after-work errands were closer to the mark. The supermarket.

And so, I went to the Walmart Supercenter to find things that go unnoticed.

It seems like an idiotic idea—who can fail to notice things in a two-hundred-thousand-square-foot warehouse expressly built to house them? But the parable of the lost keys shows that it takes concerned effort to see anything in earnest. And if seeing things is the first step to making use of them as playgrounds, then there's no better

place to exercise that muscle than in the megachurch of cut-rate consumerism. So, playing big-box archaeologist, I excavated specimens
from every aisle, as if for a natural history molded from proteins and
plastics. Among them:

> Smithfield half hams
> HEET antifreeze and water remover
> cheeseburger-flavored Pringles
> Disney *Wreck-It Ralph* Taffyta Muttonfudge action figure
> *Chicken Dinners* magazine
> Bagel-fuls strawberry filling and cream cheese frozen bagel
>   snacks
> live bait
> Optimus Prime piñatas
> Old Spice deodorant, Wolfthorn scent
> 8" spiral ground anchors
> One Direction duct tape
> Assurance adult diapers, boxer shorts style
> John Travolta's *Face/Off* on Blu-ray
> Hot Wheels Ozzy Osbourne die-cast Anglia panel truck
> Shaggy pink and shaggy silver seatbelt shoulder pads
> Heinz Tomato Ketchup, 114-ounce pour, store, and pump
>   jug

There were hundreds, thousands more. I tried to take a page from
my daughter's book: as I recorded a subset of them, momentarily giving each one all of my attention, I invited my brain to feel admiration
for them.

Doing so made me realize something. The fear of ordinary life
isn't limited to nuisances like lost keys. It's everywhere. It's total.
We're surrounded by so much that we either ignore or lament. It feels
harder and harder to tolerate things these days, let alone to admire

them. I don't just mean "things" in the abstract, "How are things?" sort of way, either. Rather, all the stuff, ideas, sensations, consumer goods, problems, people, creatures—everything in whose presence we find ourselves, whether we choose their company or not.

Because these matters are unremarkable, we pay them little heed. Choosing a configuration of ketchup packaging or a scent of deodorant isn't like choosing a college or a career or a spouse or a Sunday recreation. But if you stop and think about it, we probably spend as much or more of our time choosing and using mundane things like condiments and personal-care products than we do planning and executing our more dramatic affairs. These ordinary objects might seem like detritus we can simply ignore, until we realize that the Heinz and the Pringles represent invitations to play no less than keys or shopping trips, not to mention weightier matters like work and love and parenthood.

There's no doubt that our lives are punctuated by important, course-setting events like graduations and marriages and promotions and vacations and the like, but we do most of our living in between all those moments. We make coffee, we mow the lawn, we triage e-mail, we sit in traffic, we load and empty the dishwasher. Individually, we don't notice these small trials much, but over time they add up. In America, my fellow citizens currently live on average to be almost 78.7 years old. According to data derived from the US Bureau of Labor Statistics, before our time on Earth comes to an end each of us will have spent about 4.3 years driving, 9 years watching television, and on average about a year cleaning house (women spend about twice as much time cleaning as men do).[2]

In popular media reports on time-use surveys like these, the results are often presented in a tragic light. Can you believe that we spend so much time piloting our automobiles and our vacuum cleaners? What a waste. But what if, instead of assuming that things like deodorant and dishwashers were just noise, distractions we have to suffer to get to the good stuff, we treated them instead as sources of potential intrigue, delight, and joy—as potential playgrounds similar to careers, vacations, or relationships? How would we go about doing so, and how would we have to change our attitude about the world to make it stick?

PHYSICAL EDUCATION

The things to which we attend and the way we do so change us. The Italian writer and filmmaker Pier Paolo Pasolini once described this phenomenon as a different kind of "physical education" than that phrase normally suggests, one directed not toward the body, but away from it.

> The education given to a boy by things, by objects, by physical reality—in other words, the material phenomena of his social condition—makes that boy corporeally what he is and what he will be all his life. What has to be educated is his flesh as the mold of his spirit. Social condition is recognizable in the flesh of an individual . . . because he has been physically shaped by the education, the physical education, of the matter from which his world is made.[3]

When we hear sentiments like this, we like to think that *things* means noble tools, devices that fashion loutish boys into honorable men. Soccer balls and table saws, bicycles and Super 8 cameras. But just as much, or more perhaps, the sentiment refers to retail chaff like frosted honeybuns, Size Matters car emblems, Berry Fusion flavor Tums Smoothies, Zynga CityVille Monopoly. Even losing our keys changes us. A page later, Pasolini also writes, "Nothing forces one to look at things like making a film."[4] He might overstate matters: visiting a Walmart is also sufficient.

I don't mean to romanticize the big-box store. Walmart is also revolting, full of repellent, stupid products manufactured for no reason or for bad reasons, assembled thanks to the expenditure of petroleum byproducts for molded plastic blister packs or for diesel fuel to power dock cranes, container ships, and tractor trailers. Walmart causes precarious jobs, outsourced manufacture, and the dismantling of local business through global leverage. Walmart is gross and grey and sticky, like a giant Edward Kienholz sculpture that dispenses Kraft Lunchables.

But that's not the whole story, either. Trudging through the aisles, looking for nothing in particular but instead allowing anything to capture possible interest—this experience is disorienting and uncommon. I left Walmart with a full brain. The Hard Candy nail polish jars, the Dora the Explorer Cherry Berry Bubble Bath, the Whitewheat "healthy white" bread—isn't it possible to find such things deplorable and mawkish, while also being awestruck by their existence, individually and together? Healthy white bread! Can it be a ghastly foodstuff *and* a brilliant marketing ploy? Can it also be delicious, even, in the way soft, empty white bread always is?

God help me for admitting it, but I enjoyed circulating through the Atlanta Walmart more than I enjoy visits to the Atlanta High Museum of Art—my hometown mausoleum for culture rather than consumption. The High Museum is no less a monstrosity than Walmart, thanks to Richard Meier's overbearing architecture—meandering, geometric lines in all white on white. It reflects so much sun that it blinds visitors to any art they might later encounter. Long after Andy Warhol turned the museum into a general store by transforming ordinary soup cans and Brillo boxes into rarified masterpieces, Walmart has turned the big-box store into a museum, an archive of the immediate present, a bible of prosaic miracles.

The philosopher Immanuel Kant described two kinds of sublimity, which he named the mathematical and dynamic.[5] The first describes the overwhelm of immensity in size, the second the overwhelm of intensity of force. Walmart is sublime in both senses—an overwhelming collection of entities that somehow, improbably, all made it here, and the terrifying sensation of physical power such an enormous warehouse erects over us.

WAIT, SERIOUSLY?

Isn't this kind of seeing what Wallace's fish are missing? Seeing the world for what it is. Acknowledging that the Walmart is water as much as anything. But we fear such an admission. We fear it will make us seem crass or lowbrow, for example, or that it will reveal the pretense of our supposedly highbrow discernment. We fear that it implicitly endorses all the perils of global business, or else that it insults the honest people for whom Walmart is an oasis, or even that it disparages those displaced and abused by Walmart's rise. Just entering a Walmart can incite so many fears. Sometimes we fear them all at once, one fear competing with another for position, like Wallace imagining the worst-possible scenario for the lady in the checkout.

It's become impossible to do something like praising big-box stores without risking a descent into *irony*, the rhetorical strategy of saying (or doing) one thing but meaning another. I can almost hear you thinking it, in fact: *does this guy really think discount stores trump Dalí, or is he putting us on to make a point?* The critic Roy Christopher has called irony "the most abused trope of our time, a 'get out of judgment free' card, an escape route, an exit strategy."[6] It is a prevailing aesthetic in popular culture, and given that bohemianism has always occupied the flip side of privilege, it's particularly considered such among the well-(enough)-to-do who have the luxury to choose. As the Princeton professor Christy Wampole put it in an infamous takedown of "hipster" irony culture published in the *New York Times* in 2012, "For the well-educated and financially secure, irony functions as a kind of credit card you never have to pay back. In other words, the hipster can frivolously invest in sham social capital without ever paying back one sincere dime. He doesn't own anything he possesses."[7]

But if irony is an escape route, from what does it allow escape? Wampole's answer is: "from earnestness," but that's not exactly right. Rather, irony is an escape from *having to choose* between earnestness and disdain, themselves misguided efforts to deal with fear and boredom. Wearing a trucker cap or drinking PBR or instagramming a Waffle House breakfast might or might not be celebrations of those subjects, and they might or might not be send-ups of those things. Not knowing and not being able to know is the meaning of irony today. Not saying or doing the opposite of what one means, but refusing to reveal whether or not one really means not to mean it.

Like fomites gather infectious organisms to transmit from skin to clothes to doorknobs to other skin, objects now gather irony. It sticks to everything. Fractal, irony reproduces itself within, below, around, and above itself. Eventually, the ironic actor doesn't even know whether she is earnest or contemptuous. Irony becomes unstoppable, devouring everything it touches, until all signal gives way to noise, or worse, an unknowing hum that could equally serve as signal or as noise. Once ironized, things can safely disappear into the background, unseen, unthought, unused. Isn't this what water is to fish? A thing that is always there, that needs to be there for breath and motion, but that goes unseen. Imagine if the fish rejected the water, distancing themselves from it as if they were just *above it all*. How would that end?

...

IRONY AFFLICTS US all. It is the great error of our age, the result of the refinement of fear, boredom, and distance. Declaring irony an affectation of "hipsters," as Wampole does, is unfair and unhelpful, but hipster irony offers a caricature of the phenomenon that can help us see it clearly. On Craigslist, a self-declared "failed hipster" seeks to sell his $1,600 fixie bike. "I tried so hard," he writes. "I dated a girl from Portland. I criticized cheese. I burned and singed my forearms just to make it look like I was going to culinary school."[8] The ad cannot be real, but it also cannot be fake—the man really does have a bike to sell. (Unless, of course, he doesn't.) And so, it refuses to choose. One does not buy the bike but the idea of the bike. Better: pondering the idea of buying the bike takes the place of watching sitcoms or knitting.

Irony has become ubiquitous partly because its neither/nor logic spreads so rapidly, infecting everything. Christy Wampole was even forced to respond to the assumption that her *Times* piece must have been an ironic performance. "Some people seem to think that the whole piece is ironic," she told *New York Magazine*, before clarifying, "I was actually quite sincere in my argument."[9] For its part, the magazine ratcheted up the ironic stakes by avoiding taking a position on the issue and instead running a story about the very idea of confusing a rejoinder of irony for irony itself, ha-ha-ha.

This kind of irony erupts daily on the Internet. One example among the endless options: an image of the somewhat washed-up former child star Macaulay Culkin wearing a T-shirt of the much more successful former child star Ryan Gosling wearing a T-shirt of Macaulay Culkin, in response to Ryan Gosling having worn a shirt of Macaulay Culkin. And then, of course, the inevitable photoshop of Ryan Gosling wearing a T-shirt of Macaulay Culkin wearing a T-shirt of Ryan Gosling wearing a T-shirt of Macaulay Culkin.

It's remarkably easy to produce this kind of irony, far harder than it is to accept Pasolini's invitation to incorporate external things into internal experience. As an illustration, I offer you the Tumblr blog I created of all of the Walmart products I mentioned above and a hundred more, which you can visit at supercenter.tumblr.com. Making such a thing doesn't celebrate or sneer at Walmart or its products, so much as it suspends them in an inoculating gel, perpetually frozen between the vats of sincerity and cynicism. You can partake, too, by reblogging or tweeting or otherwise caging and rereleasing my grey absorbent pile universal toilet-lid cover or nabi Bumper "I only dropped it a little bit" Kindle Fire case.

In a mutation of an Internet adage known as Poe's Law ("without context, it's impossible to tell the difference between extremism and its parody"), paean and burlesque have become indistinguishable, too. Fandom and travesty, love and hatred average out into an indistinct, pink-grey sludge. Irony is like a wink from an android. You think you know what it means, until you realize the signal you took for meaning emanates from a source for which meaning is meaningless. Unless it isn't.

THE AGE OF IRONY

How did irony become the currency of our age? Perhaps as a defense against the feeling that anything could go wrong at any moment. That you might end up stuck with your dad in a mall on a perfectly good afternoon, or that you might become lodged behind a big SUV in traffic on the way home from the supermarket after work, or that you might accidentally destroy the expensive sod you installed mere weeks after having laid it. Irony is cultured by the (justified!) fear that anything whatsoever might turn on us. We are so concerned that something might surprise us with disappointing or calamitous results that we resist engaging with it as a protection. We attempt to create a buffer atop the real world that protects us from this risk. Irony is like that plastic cover on your grandmother's sofa: in protecting the furniture from hypothetical accident, it also removes its upholstery from possible experience.

Irony is the *opposite* of a playground. Rather than embracing, creating, or otherwise accepting the ultimate existential preposterousness of the world and working with it nevertheless, irony takes the first step—drawing the boundary, encircling the materials with which one might then produce novel experience—and then . . . it stops with a chuckle and a sneer. Rather than accept either the protected or the exposed state of the plastic sofa cover, irony celebrates the buffer—the plastic—as an alternative. Where grandma deployed the plastic cover out of paranoia that some mishap might befall a piece of furniture meant to last a lifetime, irony deploys it for the sake of experiencing the cover as an alternative to the cheap, crappy sofa that doesn't warrant protection in the first place. Irony sells plastic sofa covers from the back of a truck in the IKEA parking lot. It deceives both paranoia and protection in order to install a new order grounded in the rejection of both.

In fact, irony once even meant *deception*, or feigned ignorance. In its Greek origins, irony (*eirōneia*) is named after the Greek comic character type *eiron*, who deploys wit and understatement to overcome his adversaries.[10] This dissimulation is common in the writings of Plato, where Socrates feigns ignorance to draw out the responses of his interlocutors so that he can systematically dismantle them from the position of wisdom that appears when he shows his hand.

Dramatic irony, as found in the tragedies of Sophocles, is similar but less tactical. In *Antigone* or *Oedipus the King*, the tragic hero is deceived rather than the interlocutor—deceived by confidently misunderstanding his or her situation. The Oracle at Delphi tells Oedipus that he is destined to marry his mother and murder his father, but Oedipus is mistaken about the true identity of both parents. So he proceeds with the confidence and hubris required of a tragic hero, only to be undone by his own actions, which he thought he was pursuing precisely to avoid the prophecy. Dramatic irony is covering the sofa with plastic only to discover later that the plastic is trapping moisture underneath, ruining the upholstery with fungus.

...

THEN THERE'S THE kind of irony we all learned about in the 1990s from the infamous Alanis Morissette song. In this case, irony entails any number of things, from coincidence to disappointment to bad timing. The song is now perhaps more famous for supposedly misconstruing irony than it is for anything else—it's ironic, isn't it, wink-wink?

Actually, the song "Ironic" does contain one viable, traditional example of dramatic irony, the bit about Mr. Play It Safe. He's the man who fears flying, but finally works up the courage to take a flight only to meet his doom: "And as the plane crashed down he thought, Well, isn't this nice?"

It *is* ironic to fear flying, to finally overcome that fear, and to discover that the fear was in fact justified, even if unlikely. It's furthermore ironic in that textbook "say what you don't mean" fashion to exclaim, "Well, isn't this nice," as your plane tumbles fatefully to Earth (that's called *verbal irony*).

Alanis's other quips are mostly failed examples of the domesticated version of dramatic irony, sometimes called *situational irony*. In this sort of irony we imagine ourselves to be observing the world as if it were a fiction. O. Henry's famous short story "The Gift of the Magi" (about a young couple's ironic self-defeat in mutual gifting) embodies situational irony, but rain on your wedding day, or a free ride when you've already paid, or a traffic jam when you're already late—these similarly sized misfortunes do not. They are only unlucky, even if unexpected, circumstances.

Shortly before Alanis took on irony on the airwaves, David Foster Wallace addressed the topic in an influential essay on the impact of television on contemporary literature. Wallace celebrated what he saw as rebellious irony in the postmodern fiction of Don DeLillo and Thomas Pynchon and William Gaddis, and he lamented the resulting "enfeebling" that accompanied its loss—"the voice of the trapped who have come to enjoy their cage."[11] As in the plastic-wrapped sofa, this new, devitalized irony celebrates rather than critiques that cage.

DeLillo in particular serves as a clarion call against the helplessness that Wallace argues literature had inherited from television. As evidence for irony's once vigorous critical capacity, Wallace cites De-Lillo's 1985 novel *White Noise*. There's a scene in which two professors, Murray and Jack, visit the most photographed barn in America. Wallace calls it the "metastasis of watching," in which "not only are people watching a barn whose only claim to fame is as an object of watching, but the pop scholar Murray is watching people watch a barn, and his friend Jack is watching Murray watch the watching, and we readers are pretty obviously watching Jack the narrator watch Jack watching, etc."[12] The destructive absurdity of the situation is supposed to be obvious. Maybe it once was, even.

But three decades after DeLillo's heyday and two beyond Wallace's, here in the age of irony, this sort of supposedly rebellious irony hardly seems rebellious. It feels like business as usual. Take Alanis herself. It won't take too much googling for you to find someone proposing, maybe even ironically, that Alanis deliberately wrote "Ironic" because it contained no actual irony, but only, you know, ironic nonirony irony. Don't you think? There's that buffer again, irony as a replacement for both engagement and rejection.

But soon enough, irony's buffer only reminds us of what lies on either side of it: either the fear of risk or the risk of contact. DeLillo's metawatching was supposed to be a warning, not an inspiration. But now we can't get enough of sofa-wrapping. We eat it up. Even Alanis is in on it! Twenty years after the song's 1995 release, she and James Corden performed an "updated" version of "Ironic" on the *Late Late Show*: "It's like singing 'Ironic,'" they intoned, "but there are no ironies." Irony is unable to contend with the simultaneity of risk and contact, so it constructs a membrane between them, which, tragically, makes its object inaccessible to experience. (Save for the experience of inaccessibility, of course.)

The young husband in "The Gift of the Magi" sells his heirloom pocket watch to buy combs for his wife's long, carefully kept hair, only to discover that she has sold her hair to a wigmaker to buy him a watch chain. Today's irony is that of the husband who buys the $5 digital Casio in order to ensure that it never has enough value to inspire pawning anyway, while simultaneously wearing the watch religiously as a deliberate retro-fashion rejection of the *very idea* of value. As Wallace put it, "The reason why our pervasive cultural irony is at once so powerful and so unsatisfying is that an ironist is impossible to pin down."

Irony is the escape from having to choose between earnestness and contempt. Rather than embracing an object like a pocket watch earnestly as an heirloom or a craft object or a timepiece, and rather than rejecting it as an affectation or a conceit or a pretention of conspicuous consumption, irony demands that both and neither be the case. Who needs the Casio when you've got your iPhone to tell the time anyway?

IRONOIA

We still want to skip over the proverbial sidewalk cracks. We still want to fashion the playgrounds that would lead us to novel experience. But we've convinced ourselves that we can't or shouldn't. That it's not worthwhile to embrace things for what they are, or that it *is* worthwhile but ultimately futile.

This anxiety expresses itself constantly, our minds flip-flopping between heartfelt commitment and scornful disdain. The proud, green lawn might transform overnight into an expensive embarrassment; easier to avoid even wanting to own a house you can't afford anyway. Wooing the cute somebody at work might require negotiating mutually conflicting social circumstances; easier to swipe right on Tinder.

Earnestness and contempt are close neighbors; the thinnest membrane exists between them. When pushed to their limits, each overheats and transforms into the other. At its extreme, sincerity becomes mawkish, sentimental, cheesy. Your great-aunt's Hummel figurines, or your boyfriend's exasperating, suffocating pandering, or your devout father-in-law's constant zealous devotion. Such matters force out all feeling, converting earnest interest into bitter contempt.

But for its part, when pushed far enough, contempt overheats and becomes so appalling that it reverses into a new sincerity: your great-uncle's outspoken and constant disdain for your aunt's figurine collection, or your cruel plans for the emotional destruction of your hapless boyfriend, or your silent revulsion during Thanksgiving grace at your in-laws. The absence of feeling generates a surplus of feeling, and in the face of contempt one can't help but feel newly earnest. Once something becomes thing enough to evoke either sincerity or contempt, it ratchets up into a high-pitched, fluorescent hum between the two. And like the throbbing fluorescent lamp, that humming produces headaches. The oscillating anxiety of irony's undecidable earnestness or contempt—not either one or the other, but never being able to discern the one from the other—is a kind of madness.

Ironoia, we might call it.

If paranoia is the mistrust of people, ironoia is the mistrust of things. It is our commonest but most infrequently diagnosed condition. And it comes with an equally common folk cure: to seek escape, to recede further away from a thing rather than to stop and attend to one by circumscribing it in the zone of attention that might create a playground. Zooming out a level with irony is far easier than reconciling the conflict between sincerity and disdain so as to reconnect with the world we miss by wavering endlessly between them.

The many tools we have for ingesting and representing objects, concepts, and ideas makes this move more effortless—particularly life online, where the cost of accessing, producing, and distributing them is effectively reduced to nothing. Making a T-shirt already requires limited effort today, but photoshopping and posting one online takes almost none. "It's not funny when no one knows the person on your T-shirt, or you for that matter," writes Ian Crouch in the *New Yorker* about the Macaulay Culkin/Ryan Gosling T-shirts.[13] Another, gentler call for earnestness. But, of course, the moment anyone says anything of the sort, they've already gotten it wrong. The purpose of such shirts is both to embrace *and* to reject identification—and never to admit which move is primary.

Irony is a hedge, an insurance policy against further affliction. It erects boundaries that we hope will protect us from the world by forbidding it access to us. Or, differently put, we double the world, creating a safe, fictional version atop the real one—whether in our heads or on the Internet. We believe that interacting with this fictional copy of the world will save us from both pain and boredom.

But instead, once the distinction between fiction and reality breaks down, as it always does, we end up rewrapping the proverbial sofa in a new layer of plastic. And on and on and on we go, wrapping and rewrapping our lives and everything within them in increasingly more futile layers of protective covering, all the while knowing that the last one failed to protect us even more than the one before. Irony is an asymptotic death march into nihilism, one that can legitimately claim it averts futility by never quite getting there.

■■■

A WHILE BACK, a game designer friend of mine named Phil Fish made a plea on Twitter, "Hey bloggers, no more 'blank rebuilt in Minecraft' posts, please. We get it. You can make things in Minecraft. Thanks." Fish was referring to the popular online game Minecraft, in which players hunt for resources that are used to construct models and apparatuses with the game's characteristic, cubical visual style. The Internet being what it is, given such tools extreme fans do insane things, like elaborately reconstructing the city King's Landing from *Game of Thrones* using nothing but this square matter mined from Minecraft.

Seeing Fish's tweet, an enterprising ironoiac recreated the form of the embedded tweet itself inside Minecraft, a fact that the tech blog VentureBeat then dutifully blogged about, thus completing not one but two cycles of an ironoia self-treatment the environmental philosopher Timothy Morton names "anything you can do I can do meta."[14] In a futile attempt to prevent further metastasis, the blogger concluded his post with the line, "Yes, we're fully aware of the irony of this post."[15]

But rather than satisfying anyone, such a provocation only further irritated the ironoiac itch. Fish tweeted a link to the blog post covering the Minecraft construction of a model of Fish's tweet protesting blog posts about Minecraft constructions, which one of his followers one-upped by observing the fact that Fish had in fact "tweeted about somebody blogging about somebody making [his] tweet about Minecraft in Minecraft." Another chimed in, "How long 'til someone recreates that blog post in Minecraft?" Each step represents an attempt to overcome the absurdity of the last by fixing it in a new voice, even though each ironic gesture was evanescent, quickly replaced by yet another layer of buffer from yet another desperate ironoiac.

Why do we do it, then? Today, satisfaction is more elusive than ever. In part, the precarity of life after the 2008 global financial collapse and the Great Recession that followed it (and whose effects still linger) makes every transaction with the world feel suspect and risky.

We fear that things might turn on us, because we have good evidence that they can, and do.

But this condition has expanded from an economic to a cultural one. Even enormous material wealth doesn't stave off the spread of boredom. In late 2014, Minecraft creator Markus "Notch" Persson sold his company to Microsoft for $2.5 billion. Notch published a depressive justification for his desire to recede from public life thanks to the impossibility of satisfying the onslaught of demands from his customers and fans—another thing that can turn on you, it turns out. Then he bought a $70 million Beverly Hills mansion, along with all the furnishings, accessories, art, even the cases of champagne and tequila, even the ultraluxury vehicles the real estate speculator who built the place had installed within its sprawling garage for staging. Notch, the man who made a blank canvas world in which you could make anything, used the spoils to buy a prepackaged, off-the-shelf billionaire's life. As for his fans, undeterred, they dutifully reconstructed a version of the $70 million mansion in Minecraft.[16]

It's an addict's logic: only one more hand, only one more hit, then I'll be satisfied. Then I can stop. But, of course, that's not how addiction works. With every repetition, the effect of a compulsion reduces, requiring even more stimulation to produce formerly intoxicating results. Such is today's irony: a narcotic whose potency decays every time it is used. When deployed, both sincerity and contempt short-circuit, but only briefly. No wonder we're always bored. No wonder we're ever more afraid.

"Anything you can do I can do meta" feels like a salve for ironoia because it seems to inoculate something against further enclosure. It attempts to break the cycle of eternal, plasticized rewrapping. "Finally my boredom will cease," we think. But rather than working with the playground that an initial boundary had established—rendering a piece of Internet media in Minecraft cubes, say, or evaluating the rhetorical tropes in Alanis's supposed ironies—irony acknowledges then rejects that opportunity, fashioning another one atop it as revenge for having been subjected to the very idea of a bounded space in which playful experimentation is possible. You blog about tweeting about minecrafting a tweet, or you wink coyly through an ironic "Ironic" on late night television.

"Going meta" forgets the self-replicating, fractal structure of irony. A fractal is a pattern that recurs in the same fashion at smaller and larger scales. You've probably seen fractals like the Mandelbrot set as computer simulations: zoom in or out and the same complex pattern recurs infinitely. The mathematics of fractals are often used in computer graphics to generate natural environments like mountains or rivers. Some even claim that the very fabric of the universe is fashioned from something akin to fractal math, and that enormous complexity can be described with relatively simple "instructions."

But ironoiacs mistake the beauty of the fractal with the banality of any single view it produces. The *concept* of the fractal is sublime in its elegant simplicity, but individual views on one are just corny: yet another computer image you've seen a million times before. Meanwhile, a mountain or a riverbed is sublime not because it resembles the results of fractal math, but because it is a *particular* mountain or riverbed. The fractal is like grandma's proverbial davenport in infinite regress, forever covered and recovered in plastic. Zooming out or in no longer matters when everything is always the same.

Even so, we still believe that something different lies on the other side of our ironic acts. It's why films like Christopher Nolan's *Inception*, with its interlocking dreams within dreams, and apocalyptic utopias like Ray Kurzweil's singularity—the idea that soon humans will be able to upload and live forever inside computers—are so appealing today. If only I could get out of this reality and into another, then I'd be content, we think, but here, nothing is sufficient. It's a nightmare perversion of a cruel children's playground game. "I'm just kidding for real," you declare, just begging me to wonder, *for real* for real?

To combat the ever-decaying potency of irony, going meta simulates the embrace of a playground, while actually signaling its predestined insufficiency. Today, we 'scare-quote' almost everything, even if we're not doing so typographically. Sometimes explicitly, while our little fingers waggle around our heads to cast suspicion or doubt, but more often implicitly, by default. What could possibly succeed at drawing out our attention without having ulterior motives? Ironoia is different from paranoia, because the ironoiac can make no appeals to the conspiracy theories of the paranoiac. While one could imagine a cabal of industrial pig farmers all secretly colluding against the paranoid, no dark secrets remain within the supermarket hams themselves. The false deviance is always put there by some agent, the persecution wrought through intermediaries like a lowly Smithfield half ham.

The ironoiac's problem is different. Objects do not represent the latent, devious plans of an organized system of persecution. Rather, they embody all the things that could possibly go wrong once an object is put to use. Paranoia is a condition of scarcity—the world's repletion having been focused, winnowed down into a weapon of persecution. But ironoia is a condition of surplus: with so many options, so many possible products, activities, or companions, with so many possible ways of living, loving, dreaming, or suffering, what if you make the wrong choice? Despite what Christy Wampole would have you think, irony is not a condition of narcissism, but of its opposite. Ironoia is the result of oversupply, of abundance. The world is so generous, so full of choice and variety, that any individual element must be held at arm's length in case something better comes along.

NOSTALGIA

One way to look for something better is to turn toward the past, where meaning and action seemed more facile and forgiving. It's no surprise, then, that irony often overdetermines a troubled relationship to history. It is nostalgic, rekindling a burnt ember from another time without ever setting it alight.

Today we understand *nostalgia* as a sentimental longing for the past. But the term first appeared in the late seventeenth century, in reference to Swiss mercenaries, a sense that reverberated through Romantic literature.[17] In its original meaning nostalgia referred to *place* rather than time: *nostos algos*, a homecoming pain, or what we now call *homesickness*. The difference between nostalgia as a sickness for another place rather than another time is substantial. Physical traversal to another place is possible, while temporal traversal to another time is not. Nostalgia in the modern sense of the term is thus always frustrated, always impossible.

This frustration makes nostalgia a perfect match for irony, because it ensures that objects remain unworkable, imprisoned. We try to reinvigorate the hollow present with the accouterments of the past, which are romanticized as certain bearers of meaning. Fixie bikes and vinyl records offer easy targets, but the craze for retro nostalgia runs deeper: the return of old toys and films and the rebooting of lost television series or video games commercialize the nostalgia normally associated with hipsterdom.

Unlike the ironoiac or the nostalgic, the player chooses presence over distance. He keeps eager watch for something new even in the most familiar situation: a new opportunity to draw that imaginary circle that fashions a playground, or a new way of comprehending the contents of one that fills her attention. The playground turns Zen mindfulness upside down, encouraging one to be *worldful* instead, attentive to the status and complexity of all the possible subjects of attention and interaction beyond one's head.

In an earlier book on the philosophy of things, I suggested the word *ontography* as a term for taking partial record of the universe, like fashioning a register or a bestiary in order to draw attention to a diversity we often overlook.[18] My tiny archaeology of Walmart offers a concrete example. But my favorite ontographer is Stephen Shore, a former Warhol's Factory artist who popularized color fine art photography. I'm particularly fond of Shore's collection *Uncommon Places*, photos made during his travels across North America in the 1970s. The images are remarkable, partly for their subject matter: motels, diners, street corners, parking lots, and other places that, as the collection's title reminds us, are uncommon because we overlook them, not because we fail to encounter them. But these photographs also work as extraordinary ontographs thanks to their formal properties. Captured on 8x10 film plates in view cameras, Shore's images exhibit sensational detail.

But Shore's photographs are also time machines. It's almost impossible not to view these images nostalgically, noticing all the details that, to our contemporary eyes, exist out of time. The blue and white body design of the third generation Chevrolet Cheyenne full-size pickup; the ashtray and cathode ray tube television of a Holiday Inn room long lost to history; the outmoded chain-link lot barricades and the sunburst cinderblock swimming pool enclosure of a motel parking lot. The objects in Shore's images become more visible to the contemporary eye thanks to their outmodedness.

It's a common aesthetic today. Matthew Weiner made his mid-century ad-man television drama *Mad Men* as much a repository of miscellaneous artifacts from the 1960s as a drama about human beings and their failings. PR images of its empty sets could easily be mistaken for Shore's photographs: the lost drama of table lamps, of nightstand clock radios, of draperies. And thanks to *Mad Men's* success, the aesthetic of detailed, in-the-background nostalgia has spread to other period pieces of the not-too-distant past. The FX series *The Americans*, a show about two Soviet intelligence officers posing as American suburbanites in the early 1980s, deals with kitchens as much as it does the KGB: the terrible trim on refrigerators, the avocado mixing bowls, the goldenrod plastic pitchers.

When we view these shows and photographs, we achieve something improbable. We allow things to coexist with us—or at least, with our perceptions of our possible selves—rather than immediately rejecting them as potential threats or scoffing at them as instruments of boredom.

But we give up something for the privilege: we can never commune with these entities. They are partly fictional, for one, but more than that they are lost to time, gone forever, out of reach. It is easier to rhapsodize past objects because they make no threat to disappear or to fail to yield us solace. Things in the present offer no such warranty. Nostalgia cannot orient to a world we can touch, but only to one that is out of reach, a world to which no further responsibility is necessary or even possible. And so nostalgia, almost a mechanism for making playgrounds out of the ordinary world, suffocates under irony's pillow. Wallace was wrong to think that irony was ever rebellious and virtuous. Such a feeling is just another nostalgia, another thing to be ironic about.

■■■

IN THE MID-TWENTIETH century, when the real mad men were downing whiskey in real skinny ties, conceptual artists working in the wake of the avant-garde had made a similar gambit. In 1961, the artist Robert Rauschenberg was invited to create a work for an exhibition of portraits of Iris Clert, the proprietress of a successful Parisian gallery that bore her name. Rauschenberg's submission was a telegram sent to the gallery, bearing the message, "This is a portrait of Iris Clert if I say so." The same year, Piero Manzoni constructed a work comprised of ninety sealed tin cans, each emblazoned with a label reading "Artist's Shit." The cans supposedly contained the feces of the artist, but to open one to determine the truth would destroy it, thus suspending the artwork between performance and prank, a Schrödinger's litter box. In truth, works like Rauschenberg's and Manzoni's didn't reveal hypocrisies so much as create them: this is and isn't art. You can't tell. And that ambiguity isn't virtue nor genius. It's terrorism. Conceptual art as scorched earth.

Herein lies the origins—and the power—of the "hipster" irony that Wampole so despises. Objects that might have borne rich contextual meaning get reanimated not for what they are, but because they appear to posses indeterminate meaning. Potential meaning is always greater than actual meaning, and so it seems better to postpone fixing that significance as long as possible, forever maybe. Of course, when the emptiness of a divested object rubs up against the fictional promise of its future richness, the angst of ironoia sets in.

As Phil Fish's reconstructed Minecraft tweet (and any other Internet meme) show, the timescale for nostalgia need not be measured in years or decades anymore. It can activate and burn out in days or even hours. Anything that seems invested with meaning, even if temporarily, becomes a table scrap for the starving. The desire to consume it before it vanishes is too strong to ignore. And in our desperation to capture a specimen rather than to understand and use one, we suffocate it. Everyone must have a take on the latest thing.

The Tumblr blog I made from my trip to Walmart offers an example: by documenting my visit to the supercenter, I destroy the respect I brought to it as a playground for retail sublimity, even as I reincorporate its contents into a different playground, one built for garnering attention on the Internet. But I had to make the blog to demonstrate to you, dear reader, how easy and how tempting it is to wrest things for ourselves, to cover them in a fresh layer of plastic.

We have whole industries devoted to the practice, via smartphones and online services. We are all grave robbers now, even while mistaking ourselves for preservationists. Indiana Jones is no less a thief just because he wants to inhume his relics in a museum.

That's not a measure of blame, but a description of our circumstances. The source of ironic detachment comes from the fear of taking an object as something in its own right. That capacity is weakened by neoliberal austerity even as it's simultaneously strengthened by unbridled consumerism—both of which are amplified by online life. Ironism isn't really a trait of hipsters after all (itself an unfair and false category), but of the present moment of social and economic precarity that the word "hipster" euphemizes. When the possibility of home equity, health care, gainful employment, and other guarantees melt away as they have done in large part since the financial crisis of 2008, any vestige of future tangibility becomes seductive. Irony is the aesthetics of the new Gilded Age, which also happens, not coincidentally, to be the Internet age. A new trinity: Irony, Insecurity, and the Internet. All wrapped up in another *I*: the self, me, the center of the universe, the ultimate drain for all worldly runoff.

Walking around Walmart, I felt a tug, the temptation to get above those weird and stupid objects. To contain them, to make them mine. Wallace said that television was practically made for irony, but the Internet is even worse. And life in the social media age has become a new malignant addiction; an excuse to abscond with objects, people, and situations by arresting them into assets imprisoned between sincerity and contempt. Online, we become digital poachers, stealing things' souls in order to elevate ourselves above them, until we destroy those very things via incorporation or disposal. Safari spoils are made of pixels rather than ivory. The only thing better than instagramming chicken- and waffle-flavored Lay's is captioning an Instagram of chicken- and waffle-flavored Lay's with an early, definitive taste test before disposing of them in the parking lot.

SOLEMN APPRECIATION

If ironoia produces an anxiety akin to that of paranoia, then its remedy ought to come from a material counterpart to psychotherapy, a physical therapy inherited from Pasolini's version of physical education rather than from mobility remediation. Physical therapy means better connecting to the world outside ourselves, rather than from redoubling our selfish commitments.

Recall that air pocket between earnestness and contempt. Perhaps we can look there to find ways to reclaim a sense of respect for things even in the face of the ironoia that alienates us from them.

When we look at, say, the McDonald's Filet-O-Fish sandwich, it's far too mundane to recommend deliberation. Too boring. Yet another manufactured good from an untrustworthy global corporation. Yet once we "reverse engineer" the Filet-O-Fish in our own kitchens, as the writer Megan Garber explained how to do in an article in *The Atlantic*, then we can tame it, if only temporarily.[19] Once we can ironize the Filet-O-Fish, once we can do it meta, then it becomes valid thanks to having been contained.

But the irony of the ironic, DIY McSandwich is that partaking of it doesn't even require irony. The Filet-O-Fish is *already* a fitting playground all on its own. It can be encountered *even more* deeply through reconstruction. To rebuild one doesn't require holding this delight at a distance; instead, doing so offers a way to bring it even closer, to appreciate it even more by attempting to fashion one, not only to eat one. It shows us that the Filet-O-Fish is remarkable on its own merits, even if you are tempted to look down your nose at McDonald's as much as you are at Walmart.

Is there really any difference, the writer Jeb Boniakowski once asked, between highly engineered and processed foods like the kind you find at McDonald's, and molecular gastronomy, the application of food science to cooking that became popular in modernist haute cuisine establishments like elBulli and Alinea? Boniakowski draws a powerful conclusion that should be obvious in retrospect: "I've often thought that a lot of what makes crazy restaurant food taste crazy is the solemn appreciation you lend to it." But we tend to limit our indulgence of that appreciation. Boniakowski offers a delightful thought experiment to illustrate the point:

> If you put a Cheeto on a big white plate in a formal restaurant and serve it with chopsticks and say something like, "It is a cornmeal quenelle, extruded at a high speed, and so the extrusion heats the cornmeal 'polenta' and flash-cooks it, trapping air and giving it a crispy texture with a striking lightness. It is then dusted with an 'umami powder' glutamate and evaporated-dairy-solids blend." People would go nuts for that.[20]

Even as we welcome identity politics into the tiniest creases of social diversity, we still refuse to extend much of any appreciation to the material world. Instead, we mistake familiarity for a lack of authenticity. Like the cornmeal quenelle and the DIY Filet-O-Fish, the ordinary must be transformed in order to bear value. Irony defamiliarizes things, makes them momentarily curious. It fashions a playground— but then stops short. Instead of working with the material inside those borders, irony elects to reincorporate them inside a new cellophane packaging, where they can't be touched, but only observed. This distancing explains not only the desire to ironize, but also some of the mechanics.

●●●

DISTORTION IS NOTHING new in art. One could even claim that the simplest definition of art *is* distortion. Art's purpose is defamiliarization, or "enstrangement," as the early twentieth-century literary critic Viktor Shklovsky called it.[21] Poetry uses ordinary language in weird, new ways. It makes us aware of the features of words that we previously overlooked. It turns speech into playgrounds.

For example, the photo-sharing social network Instagram renders the present through the nostalgic lens of a past near and familiar enough to activate authenticity through defamiliarization. Simulated film emulsions, square print formats, lomographic lens distortions, high-contrast exposure, desaturation, color shifts, vignetting, and other effects allow a scene, an object, or a situation to be recast in a way that invites the viewer to believe in it. This access to reality is counterintuitive, because it seems as though an Instagram does just the opposite, replacing authenticity with a false veneer. But if the ontographic power of Stephen Shore's uncommon places comes partly from forcing us to see the things we previously missed, the Instagram obliges this invitation by making the familiar—even the present!—slightly unfamiliar so that we can attend to it more deliberately. By distorting reality slightly, by drawing a magic circle around it, we are able to see it more clearly. This irony is used for good, for the foundation of a playground that is—or at least, can be—used rather than cast aside.

The simple defamiliarization of the Instagram filter distortion commands our attention, setting unseen elements into relief. The point is not to see the world "as it is" (whatever that means), but to be faced with the world that we regularly look past. After all, this is the first step in my daughter's "step on a crack" game, or in my awkward negotiation with my fertilizer-burned lawn. Playgrounds are only possible when we bracket the potential boredom or trauma of the things that face us, so that their material properties can guide us to new ways of engaging them. Filtered nostalgia offers a little crutch—a tiny, gilded box of its captured contents.

That crutch offers a clue for how we can live in a world of surplus: via scaffolds that help us take things seriously for what they are, rather than reabsorbing them into the black hole of one-upmanship.

Instagram's scaffolds help us accept the arbitrary framing and contents of the scene, not as a protective casing, but as the permeable boundary of a playground.

Playgrounds are only possible when we bracket the potential boredom or trauma in the things we encounter so that their material properties can guide us to new ways of engaging them. The crappy Instagram, Shore's uncommon places, and the *Mad Men* crew's thrift-store scavenging for set decoration all do the same thing. They recast something familiar in a relatively minor way, one that yields very little to our human desires unless we quiet them through physical therapy—by working with them, by manipulating them in our heads and then our hands. By playing with them.

The dignity we invest in haute cuisine and molecular gastronomy is much like that of modern art—we do it because we get so much back from distancing ourselves from the ordinary uses of pig or of pigment. But why limit that role only to art? If crazy restaurant food does indeed taste crazy largely thanks to the solemn appreciation we lend to it, then how can we lend greater appreciation to more targets?

Wampole's cure for ironoia demands that everything come from us, from within. "Saying what you mean, meaning what you say." It's a popular notion, the neoliberal secularist's idea that now that God is dead, now that the world is made of individuals, all creation arises from the self. Wallace said the same thing twenty years ago: "All US irony is based on an implicit 'I don't really mean what I say'" before settling on the even more nth-order version: "How totally *banal* of you to ask what I mean."[22] "What passes for hip transcendence of sentiment," Wallace wrote, "is really some kind of fear of being really human."[23]

He's got the fear right, but it's not only a fear of ourselves. It's a fear of everything else, too. The problem isn't that we cannot or will not say "what we mean," whatever that means, but that we expect meaning to originate *from within us*, as if we knew who we are and what we want, as if the universe were friction free, composed of entities spilling in and out of our hands and homes and minds, unfettered by the boundaries between head and world, tire and asphalt, Bagel-ful and box. And then—*and then* that we insist on chastising those entities for failing to service us properly! "I can't even," we caption our captives. Now we're all Rauschenbergs and Manzonis and DeLillos, thanks to social media.

This burden demands that we account for everything in advance, and that we approach the world as a unified and coherent form that we have already mastered. But when we face the world, it doesn't make sense in that way. It's not there for us—not only for us, at least, even the parts that we fashioned expressly for our own ends.

I know we're concerned for ourselves. We have good reason: we are *us*, after all. But so grave is our self-concern that it risks destroying us. Ironoia isn't only a condition of twenty-something hipsters, as Wallace knew long before the rest of us did. It's the fundamental affliction of contemporary life. We are so confounded by the simultaneous presence and absurdity of a world spilling over with the surplus of being that we mistake its plenty for *our* plenty, which turns us into either sincere gluttons or contemptuous nihilists.

Such surplus is all the more reason to follow the model of my daughter, who tolerated and admired *all* things as potential playgrounds. Perhaps the problem that afflicts us is not having too many possessions (or too few), or too many choices (or too few), but in failing to know how to treat *anything* with enough respect that its existence feels like an opportunity rather than a burden. And not an opportunity because we get something back from it or because we can put it to maximally optimized use, but because we can train ourselves to approach it for exactly what it is, rather than wishing it—or we—were something different.

THREE

Fun Isn't Pleasure, It's Novelty

We think fun means enjoyment, and that we want enjoyment above all else. But we're wrong. Fun is the aftermath of deliberately manipulating a familiar situation in a new way.

WHAT DOES IT MEAN FOR SOMETHING TO BE *FUN*? IF YOU wanted to design a fun toaster, or lead a fun classroom, or advertise a fun job, or write a fun book, how would you go about it? If you wanted to find a fun appliance to buy, or a fun course to take, or a fun career to pursue, or a fun book to read, what heuristic would you choose to select one? Most of us have no idea. We don't even know what fun is, even though we claim to want it in everything. We've misunderstood *fun* to mean enjoyment without effort. Nothing has been spared the cursed attempt to "make it fun"; everything whatsoever hopes to transform itself into a delightful little morsel of sugar in your mouth.

In fact, it's in that morsel of sugar that most of us first learned about how games can make anything enjoyable, thanks to that renowned philosopher of fun: Mary Poppins. In one of the most memorable quips from the Disney film, the Victorian nanny opines on the ways "a spoonful of sugar helps the medicine go down." Despite the apparent joy this sentiment produces in the Banks children, we must not forget that they are characters scripted for a film, not real people responding to Mary Poppins's proposal through logic or experimentation. Listen to the song again closely. She only offers two examples of her theory: a robin that sings a song while fashioning a nest, and a honeybee that enjoys a sip of nectar while buzzing from bud to bud.

These are metaphors. We have no reason to believe that a robin or a bee get bored or impatient with their tasks like a child or a nanny does. And besides, Mary Poppins is either wrong or lying anyway: in truth, worker bees store nectar in a pouch called a crop, from which they regurgitate it upon returning to the hive—hardly the spoonful-of-sugar model Mary Poppins advances. So already we should be suspicious of the magical nanny.

Still, the song does a good job summarizing our current attitudes toward fun: it makes some supposedly miserable thing enjoyable. Make it something *for me*, something that I find palatable. But real fun isn't in you; it's in the world. Or better: it's in the confluence of you-and-world that a playground helps you create and see.

"A Spoonful of Sugar" offers considerably less advice about enjoyment than it appears to do. Essentially, it recommends covering over drudgery—just as the robin's song hides the boredom of nest-building, so Poppins's song hides the boredom of cleanup. Sweeping boredom under the rug works in a musical, because the cleanup itself is simplified, abstracted so it takes up only as much time as the song takes to sing. Real effort can't be so efficiently conducted, and another several choruses of "A Spoonful of Sugar" would begin to taste saccharine if matched with the actual duration of the task.

In educational technology, where the "make it fun" maxim rules, we sometimes call such advice "chocolate-covered broccoli." Slathering broccoli with chocolate doesn't transform vegetables into dessert. It mutates them into an unholy horror, something even worse than broccoli alone. Likewise, television shows, video games, and mobile apps can be diverting, but that diversion doesn't automatically make them a satisfying delivery vehicle for algebra or world history, nor a cover for chores or commuting or parenting. Like the robin and the bee the song cites, "A Spoonful of Sugar" only symbolizes the practice of making unpleasant things enjoyable; it doesn't really prescribe a method for doing so.

Actually, "A Spoonful of Sugar" tells us so little about fun that it's embarrassing we've let the song get away with it for so long. Consider the spoken verse with which Mary Poppins begins her indoctrination:

> In ev'ry job that must be done
> There is an element of fun
> You find the fun, and snap!
> The job's a game

I dare you to try to follow this advice. If an element of fun is hidden in every job, then how do you find it? Where do you look? By what process—beyond the supernatural Poppinsine "snap!"—does that job then become a game? Should I hire my own supernatural nanny? "A Spoonful of Sugar" tells us what we already know. It's a tautology: a job seems more fun if it seems more fun. So make your jobs fun so they'll be funner for you.

"Fun" has become so pervasive that you might not even notice how frequently you read, hear, or say the term. It's a sentiment as likely to be uttered by an adolescent boy on an Internet forum, as it is by Nintendo's CEO in a keynote before tens of thousands of game developers, as it is by a parent asking about a child's afternoon at a friend's, as it is by a salaryman pining for the end of the workday.

These days, everyone has a plan for making work fun, or making learning fun, or making laundry fun. In recent years, a whole industry of consultants and platform developers have even taken up this charge under the banner of "gamification"—making boring, miserable tasks like surveys, dead-end jobs, and errands "fun" through contrived rewards and meaningless metrics. Even things we might have guessed were already fun suddenly fail to satisfy—some people even want to "make sex fun," tautological though that sounds. Fun has become such an overused and confused topic that it's become almost meaningless, like a word repeated into denotative oblivion. What do we mean when we demand that something be made more fun? Do we even know?

Maybe it's the fault of game designers and critics like me. After all, "fun" rules as an assessment of artistic quality among designers of leisure and play. "A game has to be fun," we say, or, "this game isn't any fun, so let's play a different one." If anyone would know something about fun, it ought to be those who make, study, and explain games as a form of art and culture. Do we professionals even know what we have in mind when we talk about fun?

One of the oddest things about game design and game studies—the fields that explore the creation and meaning of games—is how little those of us working in them really know about our object of interest. Often, we even celebrate it: one tongue-in-cheek definition of "game" is "the historical process by which the term 'game' has been characterized and understood."[1] The philosopher Ludwig Wittgenstein argued that terms like "game" can only be defined by "family resemblance": no one set of features describes them, but only a set of shifting, overlapping properties.[2] Such hedges help keep philosophers employed, but they don't help the everyman very much.

If "fun" only indicates that a product or experience is good in some generic sense, then the term becomes a vacant sentiment. Hollywood action films and pulp-fiction novels and pop songs might sometimes be described as "fun," but such a notion would never be used as a singular measure of ultimate worth. If anything, fun carries an air of triviality. At best, a "fun book" or a "fun song" suggests nonchalance and informality. At worst, it implies superficiality and insignificance. Fun films might pass the time, but they don't win Oscars. As the game designer Raph Koster puts it, "no other artistic medium defines itself around an intended effect on the user, such as 'fun.' They all embrace a wider array of emotional impact."[3]

Perhaps games, toys, and other play activities are daft to measure their aesthetic output against a nebulous concept like fun. However, despite not quite understanding what games are or what it means to have fun, such activities nevertheless seem immensely powerful, terrifying even. On the one hand, we dismiss fun and play and games as meaningless activities, empty calories in our behavioral diets. Given a choice between "having fun" and doing just about anything else, the responsible actor would choose the anything else. On the other hand, we are obsessed with imbuing all those other activities with fun. We want work and school and chores and ordinary life to become more palatable by somehow transforming them from drudgery into pleasure. We want it both ways: to reject fun as distraction, as vice, as sloth, as insignificance, and yet also to embrace fun as a universal motivator to action.

A form of black magic is at work in games and play, the activities we most frequently attach to fun. We know they have power over people, but we can't quite characterize that power, which makes it all the more tempting to find a way to control it. This phenomenon is most obvious in big commercial games, which oscillate between dangers and opportunities in the public imagination. "What are all my students doing in Minecraft?" educators wonder. "Why can my kid lead a World of Warcraft guild but can't finish his homework?" parents lament. And all of us ask: "Why am I so addicted to Candy Crush?" We think games are powerful because they deliver fun, and we think we want fun in everything, so games must be onto something. We'd like to believe that there must be something we can extract from them and apply to our ordinary pursuits. Except, we're also terrified of games, which we see as compulsive and prurient wastes of time.

···

PART OF THE problem comes from the word *fun* itself. It's a word we use without concern for its meaning. "Have fun!" you call as a family member leaves the house, having finally found her keys. In speech, *fun* serves a perfunctory purpose more than a semantic one. More terms and phrases are perfunctory than we may realize. Much ordinary speech is automatic and functional rather than deliberate and expressive. When shopkeepers and cashiers ask, "How are you?," they don't really mean to ask how you are, but simply to say hello, to acknowledge your presence or arrival. Likewise, when you reply, "I'm fine," you're just returning the greeting.

Even seemingly momentous phrases can become deflated. The first time you tell someone "I love you" is significant because it marks a new level of intensity and commitment. But months, years, or decades later, "I love you" works more like "How are you/I'm fine"—it reinforces an expected state of affairs. It says, we're okay, everything's okay, nothing's changed. Over time, "I love you" only means anything when it's withheld, in a case when it's *not* uttered but typically would have been, after an unnoticed slight or amidst a dispute, for example.

Fun is like that. It's a placeholder more than a description. Like "How are you?," the question "Did you have fun?" is mostly a courtesy. "Yeah, we had a good time." In common parlance, it signals that everything is okay, nothing went wrong. As an aesthetic assessment, it works similarly. "This is a fun game" is akin to saying, "That was a good book." It's a generic, empty signifier that does little more than telegraph the speaker's unexamined, basic satisfaction. It may even do less than that, since, as Koster observes, other media doesn't tend to enforce a single, meaningless, and immeasurable metric for aesthetic value.

Koster's alternative might surprise you. Rather than rejecting fun as the ultimate goal for toys and games and other playthings in favor of weightier, more substantial efforts, he suggests the opposite: we ought to embrace and accelerate the pursuit of fun so that it assumes new gravity. Instead of expanding the capacity of games to achieve other emotions—say, tenderness, fear, empathy, or any of the other, deeper sensations we are accustomed to encountering through art— Koster argues that fun has greater power and capacity than we have allowed it.

What if he's right?

THE HEROISM OF ORDINARY LIFE

I'm sitting on a plane, watching passengers board, a pastime in which I find surprising delight. You can try it yourself; instead of staring at your smartphone when you reach your seat, watch and listen to the people filing past. Like a crowded subway or a supermarket checkout line, an aircraft cabin is one of a few places where material conditions apologize for and even encourage eavesdropping.

This time, two well-dressed business travelers, a man and a woman, are conversing about the indignities of modern airline travel. The man is telling a story about an unruly passenger on a prior flight. I catch only a part of their exchange, as Rollaboards greedily fill overhead space and bodies slurp into the ensuing void. "Yeah, the air marshal had to come up out of his seat, even," he recounts. His colleague responds, without missing a beat, "Ugh, that's *super fun*."

More irony, as usual. This time it's vanilla flavored: verbal irony, saying the opposite of what you mean. As the pair pass behind me into the oblivion of the fuselage, I realize how common this use of "fun" is—the ironic mention, the sneering, Morissettesque "well isn't that nice." This is fun as making-fun, as mockery or contempt. Fun as the throwing up of hands and the exhale of a heavy sigh and the slow, low shaking of the head.

In his essay on irony and television, David Foster Wallace admits that he watches TV "for fun," by which he really means for the sake of mockery: "the pleasure my generation takes from television lies in making fun of it."[4] And as programming options have expanded and television itself has merged with the Internet, that desire for fun has only further reentrenched itself as cynical laughter. In fact—and ironically—Wallace's essay on irony appears in a collection entitled *A Supposedly Fun Thing I'll Never Do Again*. The title essay, reprinted from an issue of *Harper's*, details Wallace's failure to enjoy a luxury cruise and his encouragement to his readers to adopt their own, preemptive suspicion of them too.

A coupling has developed between irony and fun. Fun has become a way of holding something at arm's length, not necessarily in order to mock it, but probably so. And yet, at the same time, *fun* still signals the supposedly desirable, life-enriching experience of pleasure that we'd seek out all the time if we were able. The "girls just wanna have fun" fun, the "everybody have fun tonight" fun. This isn't only wordplay or accident. Just as irony oscillates between earnestness and contempt, so fun swaps positions between a delightful, optimal experience (even if a vague one, but more on that in a minute), and a sneering, ironically undesirable one. Or else, as a placeholder, as a way to suspend something in between genuine appeal and clear ridicule: the contemptuous, eye-rolling adolescent "oh, *that* sounds like fun" fun.

The irony of the "super fun" experience of the air marshal coming up out of his seat to subdue the out-of-control passenger isn't only verbal—it's also situational. After all, what alternative would my fellow passenger have preferred? It's tempting to say that the absence of all experience would be ideal—just another uneventful flight that would get him or me or you from point A to point B without incident. But then, uneventful flights pose a different problem, that of boredom, flatness, and repetition. The sneering, ironic use of "fun" gives the lie to the reality of an ordinary situation like air travel: really, there's no way to satisfy us. As with dishwashers and Walmarts, commutes and Tumblrs, we want the impossible. We want air travel to be reliable and uneventful and all business, while at the same time diverting and fresh and *some* pleasure, at least.

The "fun" of a flight, it would seem, comes from the ordinary and the extraordinary all at once. Both from being left alone to sit in peace, nothing happening amidst the turbine hum, and from being distracted by conversation, movies, books, and even, occasionally, unruly passengers and the federal agents who would quell them. The fun of commercial airline flight—if such an idea doesn't immediately disqualify itself as preposterous—seems also to make equal claim to apathetic, featureless boredom and harrowing, uncomfortable distress, whether via air marshal interventions or more ordinary terrors like turbulence and delays and dead batteries.

■ ■ ■

TAKE GAMES INSTEAD of air travel. Just the other day, I watched my son play a handheld video game on the couch. As time passed, his expressions of frustration and disgust became more and more pronounced. The game was fighting hard against him, the platforms and enemies of its Super Mario Bros.–style environments resisting his attempts to advance through them. He didn't give up entirely;

in games, we have a word for that, "rage-quitting," a practice often signaled by hurling a device across the room in disgust. Rather, he persisted, as one must do, even though the only reward for persistence is the demand for more and greater perspicacity in subsequent levels. After one particularly guttural lamentation, I asked if the game was any good. "Yeah, it's fun," he replied, "it's just . . ." And he trailed off.

You can understand why. It's just *what*? I'll offer an answer on his behalf: it's just that it makes no sense to call such a ghastly, miserable experience "fun." The cognitive dissonance of such an impression wells up and halts further speech. We *want* to be ironic about fun, but then fun comes and ruins our plans by being fun *for real for real*.

You often see this friction when the subject of fun arises. After joining the social media service Twitter, where he amassed over a million followers before quitting suddenly on more than one occasion, the director, writer, and producer Joss Whedon told *Entertainment Weekly*, "What a responsibility, this is enormous work—very fun, but it really started to take up a huge amount of my head space."[5] It's an offhand remark, but such statements are often the key to deep understanding. Whedon's right: fun and work, fun and exhaustion, fun and seriousness live together as intertwined threads in a common rope. The fun, it would seem, arises from the work, from the responsibility, from the discomfort and anxiety.

The paradox of fun is this: we think fun is enjoyment, but in practice it often feels like quite the opposite. On the one hand, we'd never think to describe uncomfortable or distressing experiences as fun ones, but on the other hand, discomfort or distress often characterize the experiences we later describe as fun. A fun match of soccer might involve physical and emotional injury; a fun trip to the zoo might entail heat exhaustion and stained overalls. And yet fun doesn't *feel* like suffering either, exactly, even when it literally involves suffering. Otherwise we'd not call it fun, but hardship.

How to reconcile this paradox? For years now, critics and journalists, designers and researchers alike have relied on *flow*, the positive psychologist Mihaly Csikszentmihalyi's theory about the mentality of "optimal experience," as he calls it.[6] It's a simple notion, which helps explain its power and popularity. Flow theory contends that we perform the best when we develop an intense, engrossing focus on an experience over which we have enough control to produce and execute successfully on deliberate plans. In a diagram of mental states related to flow in his book *Finding Flow*, Csikszentmihalyi opposes flow to apathy, situating the sensation of flow between boredom and arousal, relaxation and anxiety.

This model is terrific for understanding peak performance, such as the optimized states of athletes "in the zone" or entertainers on stage. But flow theory has also been applied to just about everything else, from workplace satisfaction to education to spirituality. In nonfiction and journalistic coverage of human experiences of all kinds, one can almost count down pages until Csikszentmihalyi's name comes up, along with the requisite pronunciation guide for the uninitiated (it's CHEEK-sent-me-hy-ee, by the way).

But using flow as a model for everything poses a problem. The hyperfocused, immersive experience of flow applies well to certain deliberate activities like sport and performance, but doesn't quite jibe with more ordinary experiences. Single-minded attention might work on stage, on the pitch, or even in the classroom, but we don't spend most of our time in these places. They are exceptional.

Think of a workplace. Csikszentmihalyi argues that goals with sufficiently clear feedback to allow personal growth qualify as flow. But the ordinary, day-to-day life of the office doesn't fit so nimbly into the model of immersive attention. Even a workplace full of goals and feedback meted out at the right level to offer just enough challenge is still a workplace full of gossip and e-mails and meetings and inoperative coffee machines. Not everything is like a professional athletic contest, and not everything should be.

Flow breaks down in the office, in the airline cabin, and in front of the dishwasher. It's simply preposterous to explain ordinary activities as opportunities for "optimal experience." If every experience is optimal then none are. But worse, Csikszentmihalyi's version of optimal experience also betrays a certain conservatism. Flow isn't only an exercise in optimization, it's also one in statistical averages. It's the Goldilocks principle, where every experience must be "just right," customized to the individual who would encounter it in order to optimize his or her purchase upon it. Just as Aristotle contends that virtue is a mean between excess and deficiency, so Csikszentmihalyi holds that optimal experience can be found at the fulcrum between boredom and anxiety. By finding that mean, supposedly we can learn to ratchet up our abilities, developing new tolerances between ever more intense levels of boredom and anxiety.

Yet such is not the charge we find ourselves given before the dishwasher or in the airline fuselage or traversing the Walmart. What would an optimal airplane flight look like from the perspective of flow? Would a "fun" flight involve modest, reasonable measures of all experiences that all balance against one another? "Optimal experience" starts to seem like mixing a cocktail: one part calm observation, one part audiovisual entertainment, one part sweet and salty food sample, one part harrowing terror made flesh. Even if you could design such an experience, it would hardly make sense to call it "optimal." Indeed, perhaps the opposite is the case when it comes to air travel. A profoundly, deeply boring flight can offer challenges not easily found in other contexts. Indeed, even boredom *itself* is an experience worth taking seriously, worth encountering for what it is rather than writing it off as an ultimately fruitless, senseless sensation only good for bracketing as an unpleasantness on a psychologist's model of Experience Fluctuation.

Air travel offers one of the best opportunities for facing boredom head-on. Thanks to security and surcharges and all the rest of the dubious features of today's airline industry, airports have become dour places where one mostly corrals, queues, stands, sits, and waits. The anthropologist Marc Augé calls them *non-places*, domains of transition that don't establish enough significance to be construed as legitimate venues.[7] Augé meant this characterization negatively, as a sign of the excesses of contemporary life under capitalism—a condition he nicknames "supermodernity." Like so many sources of dissatisfaction, supermodernity assumes that all is already lost, and that airports (and supermarkets and hotel rooms—other domains that qualify as non-places) are merely faint copies of legitimate place-based experiences withheld by the cruel efficiencies of global capitalism, which forces them to be empty and self-similar.

But Stephen Shore's photographs of the "uncommon places" remind us that even Augé's supposedly Lovecraftian non-places also offer opportunities to exercise appreciation. It's far easier to see nothing of value in the concourse than to take that concourse as the given in which to find something of value, however unlikely: an out-of-the-way, quiet, plush seat in the midst of chaos; an unexpected and welcome barbeque-brisket taco; a surfeit of overpriced designer goods sold duty-free; an ebbing and flowing rush of bodies and Rollaboards.

Likewise, we can exercise appreciation in the air. I have never felt such depths of ghastly, hopeless boredom as I have on transpacific flights of fourteen hours or more. Spending that long strapped to a chair in a humming metal tube is an unnatural fate that challenges the limits of one's very humanity. It's the closest an ordinary person can come to state-sponsored torture. No quantity of movies or tasteless foodstuffs or trips down and up the aisle to and from the lavatory can fill such a burdensome amount of time. Some people can sleep on these flights, but I can't, or not for long.

In his unfinished, final novel *The Pale King*, David Foster Wallace exalts boredom, treating it as a clue to comfort. Assembled posthumously by Wallace's longtime editor, the book as it was published is hardly complete or even coherent. But the subject it dances around is clear: working a truly boring job as an IRS tax examiner, flipping pages and cross-checking documents and figures day in and day out. One of Wallace's points is that one must sacrifice individual choice and comfort to be a part of something larger. "Sometimes what's important is dull," he writes. "Sometimes it's work. Sometimes the important things aren't for your entertainment."[8] In the context of tax exams, the greater good of being a citizen of a liquid government takes priority over the moment-to-moment gratification of a supposedly passion-quenching job.

Okay, but this decree reads like moralism more than invitation. In one scene, a wizard-like substitute college instructor offers an involved and improbable soliloquy on the unsung heroism of accountants. Drudgery makes another appearance: "I know every cobble in the road you are walking. . . . To begin, . . . to hear dolorous forecasts as to the sheer drudgery of the profession you are choosing." Then he gets to the good part: "To experience commitment as the loss of options, a type of death, the death of childhood's limitless possibility, . . . This will happen, mark me." But rather than connect this "death" to terror and sorrow and decay, the substitute connects it to heroism. "Here is the truth—actual heroism receives no ovation, entertains no one. No one queues up to see it. No one is interested. . . . In fact, the less conventionally heroic or exciting or adverting or even interesting or engaging a labor appears to be, the greater its potential as an arena for actual heroism, and therefore as a denomination of joy unequaled by any."

Language like this is a novelist's rhetorical bluster, to some extent. But there's truth here, too: to experience true excitement and joy, we must first strip away all the chaff of familiarity, all the commonplace normalcy of, say, an ordinary modern jet flight—meals, movies, books, games, and lavatory visits—in order to face it for what it truly is. Joy and pleasure live beyond boredom. Under it, not before nor atop it. *Boring*, Wallace observes offhandedly, also means drilling into something, as if to find its center.

After ten hours on the transpacific flight, enough time to get from New York to Athens going the other direction, the realization sinks in: another four hours left in this aluminum prison. Here, where tolerance gives way to panic, the real work of confronting boredom sets in. Once the familiarity of something ordinary is finally, totally, utterly spent, then the novelty of facing it anew can finally start. Just as we need Stephen Shore to defamiliarize the motel or the parking lot, allowing us to see it anew rather than to look beyond it, so we need the long-haul flight to make the very idea of air travel palpable. There, in hour eleven of the Los Angeles-to-Sydney or the Atlanta-to-Seoul trek, the experience of flight begins in earnest: the terrifying and beautiful truth of 291 souls hurtling through the thin air, six miles above Earth.

Immensely distressing and even traumatic flights can do likewise. Anxiety can make the overlooked, taken-for-granted marvel of flight become newly palpable. My fellow traveler's encounter with the air marshal was worth repeating to his colleague because it is an outlier, an adventure full of nervousness and uncertainty. The experience of such anxiety would create a register for itself that would burn hot and bright in one's memory. While no one would wish to be on a flight in which a real danger to cabin, crew, and passengers were to take place, the idea of encountering one is not much different than the deliberate pursuit of other risky adventures, such as skydiving or theme-park thrill rides.

In order to find what we cannot see, we must first strip away all that we can.

FIGURE AND GROUND

Remember Wallace's fish, nestled within the water that supported them but that they could not see? Philosophers have names for this relationship between the visible and the invisible. The twentieth-century German philosopher Martin Heidegger identified two attitudes we have toward objects in the world, which he named *presence-at-hand* and *readiness-to-hand*.[9] Something is ready-to-hand when we put it to use, taking advantage of it for some end. Something is present-at-hand when it is held at a distance and revealed to be more than we'd previously thought possible. The typical Heideggerian example is a hammer. A hammer is ready-to-hand when used to drive nails. But to see the hammer as present-at-hand usually requires some disruption or calamity. A broken hammer forces you to see the tool's unexplored potential rather than remaining transfixed by its familiar actuality. Fun will require us to see the hidden potential in ordinary things so that we can put them to new uses.

The defamiliarization common to art offers another example of the distinction between use and potential. When we plow through the streets of Los Angeles, partaking of its asphalt roadways, squinting through the haze of its fog, taking care not to rear-end the Volkswagen Golf a car length in front of us, and so forth, we treat it as ready-to-hand. We make use of it, and without even noticing that we do. But to see those elements from afar, outside the context of our typical uses of them, we need something to set them in relief.

Think about what happens when you're a passenger in a car or taxi in an unfamiliar city. Suddenly, seen fresh, all the details that surround you become visible: the shape and density of trees, the noise barrier lining the highway, the shape and typography of the sign for a regional fast-food chain like Whataburger or Culver's, or even the way the forms of the vehicles weave in and out of traffic. It's all as it always is, but in ordinary circumstances you can only perceive these things as ready-to-hand—there for you to use—rather than present-at-hand—just there.

It's tempting to think that this "tool analysis" of Heidegger's distinguishes between use and nonuse: the hammer in practice and the hammer in theory as an abstraction or a formal amalgam of wood and iron. But the contemporary American philosopher Graham Harman offers a clarification: presence-at-hand shows that things are always *more* than our perceptions or uses of them reveal. For Harman, they are even more than *anything*'s perception or uses of objects reveal, not only humans. A hammer, for example, can stress the tensile strength of a nail or fall upon and crush an insect no less than it can build shipping pallets for humans to transport shrink-wrapped, blister-packed plastic gewgaws to Walmarts. Everything has greater potential than we initially suspect—or than we can ever fully know.

Marshall McLuhan, the Canadian media theorist who became a minor pop-culture celebrity in the 1960s, offers another take on the visible and invisible. McLuhan's famous adage, "the medium is the message," is but one of many versions of his insistence that media contain, reform, recombine with, and transform other media, rather than being mere transmission channels for content. And "media" for McLuhan doesn't only refer to mass media like newspapers and television and websites and apps, but to anything whatsoever that "extends the senses of man." Famously, lightbulbs count as "media" according to McLuhan, as do apartment buildings, automobiles, and classrooms.

For McLuhan, all media operate in a context. Perhaps you've seen those graphic designs that look like one or another subject depending on how your eyes focus on the scene: a vase when seen one way, but two faces when seen another. Or the image of a young woman looking away from the viewer, which when viewed again becomes an old woman looking toward us. These types of images are derived from the German theories of psychological perception known as Gestalt. While Gestalt psychology explains many different ways our eyes see distinct elements as unified wholes, images like the vase and faces are a particular sort known as *figure/ground* images. They rely on the viewer's perception of part of the subject as foreground (figure) and part as background (ground). The shift in the apparent form of the figure results from our ability to flip figure and ground, so that the vase becomes the background for the faces or vice versa.

McLuhan took the figure/ground idea and applied it to media of all kinds—which is to say, to things of all kinds.[10] He even used the same metaphor Wallace did to describe ordinary life: the fish who does not know he is in water.[11] To understand a medium like electric light or television or the Internet, it's necessary not only to understand that medium's properties (the figure) but also the contexts in which those properties become prominent (the ground).

On first blush, the figure seems like the important bit—it's the aspect of a medium we see. It's the reason we obsess about new technologies and how they are changing our lives for the better (or worse). But figure cannot exist absent a ground that makes it possible, a backdrop against which the figure emerges. And likewise, all media are made possible by the environment in which they thrive.

It's counterintuitive, but the goal of a successful medium is not to remain figure, even though the figure gets all the attention: the news stories, the nonfiction trade books, and the calls for celebration or aspersion. Why? Because figure is a curiosity, while ground is the mainstream. A mature, powerful medium shifts from novel, newfangled figure into unseen, contextualizing ground. Thus ground, the invisible component in the pair, becomes more important than figure. Radio, television, mobile phones, and the Internet offer examples of media that were once figure—new, wild, uncertain, and threatening—only to become important on a global scale once everyone had them. Ground is what we take for granted.

•••

IN CONTEMPORARY CULTURE, *fun* is perhaps the ultimate ground, the final, unseen regulator of our desires and our behavior, about which we form few theories and pose few questions. Even when *fun* appears to act as figure. Take the journalist Jennifer Senior's bestselling book *All Joy and No Fun*. The book is a lovely tour of contemporary parenting in middle-class America, with a welcome, reflective approach in an age of definitive, reductionist, and big idea prescriptions. But that title. All joy and no fun. What does it mean?

The aphorism, the reader quickly learns upon cracking the book's spine, comes from a friend of the author.[12] The idea feels intuitive and compelling: parenthood is hard work. It feels miserable at times. It involves sacrifice, doing other than one might choose. And yet it delivers delight and meaning. The book's final chapter explores the latter topic, joy, in great detail. Fun, by contrast, is never defined, let alone explored, in the book, despite its starring role in the (indisputably brilliant) title. Fun, it would seem is, well, *you know* . . . and just like my son and his video game, we allow the idea to evaporate back into the air.

While she never says so explicitly, Senior implies that fun is pleasure, whereas joy is meaning. Responding to the 2010 *New York Magazine* article that formed the basis for Senior's book, the Oberlin College psychology professor Nancy Darling helps clarify the distinction. Parenting is hard work. And for Darling, the labor of parenting—things like nagging kids to do chores, helping them with homework, encouraging them to practice instruments, etc.—are the "least pleasurable part of my interactions with them."[13] There's the lack of fun. By contrast, the more passive, ambient features of parenting—many of which result from the work of nagging and helping and encouraging—constitute the meaningful aspects of the practice. A spontaneous hug, a favorable report card, or a successful recital. And there's the joy.

The idea that fun and pleasure are intertwined seems intuitive and beyond debate. What is fun if not pleasure? An easy, passive pleasure, too, as if Nancy Darling could reap the crops of parental meaning from the comfort of the sofa. And yet, the resulting experiences *still* don't get to be pleasures, but just joys: sensations of pride in the accomplishments of one's children. There's a Yiddish word for it, *naches*, which derives from the Hebrew word for contentment.

Isn't it interesting that the hard, even miserable pleasure of playing a game is something we're willing to call *fun*, but the hard, even miserable pleasure of something far more substantive and important, like parenting, gets defined categorically as *not fun*? Perhaps joyful, perhaps even pleasurable in its ultimate ends, but always fringed with the embroidery of polite, civilized life. Why, we might ask, are we so concerned with the *feeling* that parenting (or games or airline flights or dishwashers) impart *on us*, and so unconcerned with the *nature* of raising children, or playing games, or traveling on planes, or washing dishes? What if we have so little fun not because our world is so unpleasurable, but because we've gotten fun wrong?

FOOL'S ERRANDS

Koster follows a long line of anthropologists, psychologists, and other critics who have maintained that play and games are central to human experience rather than distractions from it. The Dutch historian Johan Huizinga argues that play is the central element through which culture enacts itself. But even Huizinga, writing in 1938, isn't sure about fun. It "resists all analysis, all logical interpretation," he says.[14] Huizinga concludes that words like *fun* are thus irreducible to other terms, but this explanation fails to satisfy.

Huizinga also observes that fun seems to be a uniquely English concept, one not readily translatable into Dutch or German or French. And as it happens, the word *fun* finds its origins in the English word *foolishness*. The Middle English word *fon* that would become *fun* means "a fool," or to "make a fool of"—like we might say "don't poke fun at me" today. Less common still is *fun* as a substantive—an activity, presumably a fun one, as in the early seventeenth-century example cited in the Oxford English Dictionary—"A Hackney Coachman he did buy her, And was not this a very good Fun." Fun meant a particular type of jocularity or diversion, one meant for or done by fools.

Huizinga rejects the connection between folly and fun; play is "not foolish," he argues. But there is indeed something unreasonable and even foolish about playing games. Perhaps the problem comes not in fools and foolishness, but in construing them as pejoratives. Imprudence may characterize one aspect of the fool, the jester, and the trickster, but its flipside is *commitment*. That may sound strange: indiscretion is normally opposed to commitment. But fools have their own shrewdness: instead of toeing the line, maintaining the standard way of things, the fool asks what else is possible, then carries out even the most outlandish answer. The surprise of foolishness arises from exploration rather than from witlessness. The fool finds something new in a familiar situation.

The medieval fool or jester was a trickster, but not a reckless one. Being a fool was an actual honest-to-god job. While bumpkins or simpletons might also be called fools, as a member of the court a special status was conferred upon the jester, a status that allowed him to offer insights or observations that might otherwise be off-limits due to their excessive frankness. Being a fool was a *commitment*. The fool was expected to see life differently. Queen Elizabeth is even said to have rejoined her fools for not being critical *enough* of her reign.

The fool asks what else is possible, and then makes that alternative visible. That task takes shrewdness, not witlessness. It requires painstaking attention to detail to find something new in a familiar situation, not the freewheeling we usually think of when we think of fun. The fool teaches us that fun requires a *greater* commitment to everyday life, not a lesser one.

A friend returns from an evening out. "How was your night?" you ask her. "Fun. We had a good time," she reports. What does she mean? Even with the same friends at the same bar with the same hot wings and the same complaints about the same coworkers, the evening resulted in some new discovery. The way a particular sense of humor responded to a particular story. The way a face blanketed a new worry with a familiar gentleness. Just like "I love you," "We had fun" is compressed, condensed shorthand for poetry that goes unwritten.

Here's where Mary Poppins leads us astray. A spoonful of sugar covers over something, hides it, turns it into a lie. The sugar assumes that the situation itself is always insufficient, that its structure never holds up to scrutiny, and that it's incapable of delivering anything more than has already been thought about it. A singsong job works in a movie, but when the cold reality of a big, messy room or a long, boring flight faces us, a song and dance becomes an adornment, an affectation. And we feel guilty for that feeling because we were weaned on children's stories that taught us that meaning comes from outside a situation. That we must import or manufacture it, lest we be lazy or ungrateful.

An old aphorism about golf calls the game "a good walk spoiled." That funny quip underscores a fundamental feature of games: games make no sense, and yet we take them seriously *precisely because* they make no sense. The philosopher Bernard Suits calls that seriousness "a voluntary attempt to overcome unnecessary obstacles."[15] Golf is a desirable experience *because* it distorts space and time in order to make the player's experience of a landscape more deliberate. We seek out this deliberateness when we play.

Fun is not a feeling, it turns out. And it's certainly not the feeling of enjoying ourselves by doing exactly what we want, by making something easy or by rewarding ourselves with points, as if life is some latent version of Space Invaders that turns chores into chortles. Instead, fun is a name for deliberately manipulating a familiar situation in a new way. We can *only* have fun once we have accepted the truth of a situation and treated it for what it is. When we've agreed to suspend our disbelief in its preposterousness. Golf isn't a good walk spoiled, but a way to transform landscapes into a centuries-long hobby.

And like golf, the things we tend to find the most "fun" are not easy and sweet like the Bankses' cleanup routine. Manual transmissions and knitting are fun because they make driving and fashion *hard* rather than easy. They expose the materials of vehicles and fabrics, and they do not apologize for doing so. They make playgrounds in which gear ratios and yarn loops become materials like ceramic floor tiles or zoysiagrass or espresso. Terror is at work in real fun, the terror of facing the world as it really is, rather than covering it up with sugar.

The process of fun turns out to be work. But not work like you find in the workplace. Rather, work like you find when *working* something: working wood, or working the muscles, or working the dance floor. In each of these cases, the extraction of pleasure or enjoyment we derive does not constitute the fun of the act, but the experience of oscillating between the potential and actual properties of a set of people, things, events, and ideas outside of us. In Marshall McLuhan's terms, fun is the process of flipping figure and ground, taking something unseen and forcing it to become visible again or anew and taking something obvious and hiding it away temporarily. In Martin Heidegger's terms, fun is taking something ready-to-hand (in use, but unseen) and reframing it as present-at-hand (outside implementation, with potential). And then, after circumscribing a playground around it to set its materials in relief, imagining and deploying new actions on and with that thing.

This explanation may sound somewhat abstract and unnecessarily philosophical, but actually, it's precisely what people do when they partake in the activities they think to call fun.

Golf isn't a good walk spoiled, but an approach to encountering a designed and tended environment. A playground made of playgrounds. First, a natural environment is reformed into a manicured one, with fairways, rough, sand-and-water traps, rocks, curves and switchbacks, trees, and so forth. In this respect, the golf course is similar to the English garden, insofar as both create an invitation to pastoral traversal. The garden stops there, and that fact is its delight: the opportunity arises to watch, to think, to listen, to stroll, or to attend to nothing whatsoever.

But at the golf course, traversal is only the beginning. The ball and the clubs and the rules of the game shift the ground of the course—the actual grounds, in this case—into figure. They recast the fairways and roughs and greens as possible vantage points or obstacles for moving the ball toward the hole. New playgrounds get circumscribed as others fall away, and then spring up again anew. The golfer's own strength, experience, adeptness, and tactical wherewithal perform further oscillations of figure and ground, use and potential. So do the other material conditions that might intersect with the game: the weather, such as the effects of heat on the golfer's stamina or the effects of a prior rain on the ball's ability to rebound from a bounce on the fairway; the time of day and week, and its impact on the crowds on the greens; the familiarity of a local course, or the novelty of a new one; the languor or pressure of a particular foursome, filled with friends or family or strangers or business prospects. Instead of a good walk spoiled, golf is a walk refactored, turned in on itself like a helix, again and again.

The gyres and volutions of fun activities help explain another counterintuitive property of the phenomenon. Fun isn't a distraction or an escape from the world, but an ever deeper and more committed engagement with it. The golfer who fails to engage with the various figures of the course and its related materials, whether through incompetence or refusal, is likely to have far less fun than the player who has, through practice and attention, found reason to address those conditions as a part of the play experience.

Proponents of Csikszentmihalyian flow would add a codicil here: that the novice or the amateur's fun only arises from goals and capacities carefully balanced between anxiety and boredom. And here, the role of flow may indeed have a place: difficult courses like Pebble Beach or Torrey Pines push even competent golfers to the edge of useful action. Without the capacity to engage the materials of golf *as* materials, to work them like one works wood, the resulting experience would be inferior.

But these local minima of fun rely too heavily on human goals and accomplishments, and not enough on the unassailable nature of reality. Flow may help us optimize experience, but *viscosity* is needed to understand where such optimizations might take place. A supposedly inferior experience, like the novice golfer at Pebble Beach, is only terrible if we are unwilling to engage with boredom or anxiety in the first place, if we are so selfish as to see our performance as central to the encounter with a sublime place like Pebble Beach.

Fun is not only the delight in success, but also the panic of uncertainty, the agony of failure. It arises when figure and ground swap places and surprise us. The familiar turns strange; we no longer grasp it fully. There, facing the world's stark truth, we either throw up our hands in disgust or dread—or we persist and discover something new.

...

THIS PROCESS OF discovery that we have heretofore nicknamed *fun* is a good candidate for the physical therapy we require to overcome ironoia. Irony, remember, arises when we fail to trust things as they are, but instead hold them at arm's length for fear they will turn on us or otherwise disappoint. But in the symptom we also discovered a possible cure: treating the sensation of ironoia as a signal that something is worthy of further attention. As a sign that we've already found or constructed a playground—all we need to do is accept the invitation to make use of it.

The desire to encapsulate something in the safety of a new covering or to reject it as insufficient signals an opportunity to renew one's commitment toward it. My flight-mate's discomfort at the air marshal (the ironic not-fun "super fun") or the novice golfer's anxiety when facing the unfamiliar rituals of a foursome—these moments of distrust offer evidence that something deeper and weirder is present than each agent had previously considered possible.

And just as Wallace argued that satisfaction and joy and contribution might come from work—even boring work—as much as from entertainment, so the feelings that we might extract from a particular playground are far more numerous and more complex than the simple, easy pleasure we have mistakenly named "fun." Anxiety, uncertainty, sorrow, and even misery and horror prove compatible with fun, once we've allowed ourselves to think of *fun* as treating something for what it is rather than as a nickname for diversion or satisfaction.

A job is made fun not by turning it into a game, but by deeply and deliberately pursuing it *as a job*. Jobs are fun when their work is meaningful, when their activities matter, and when the act of conducting them can be done over and over again with increased commitment. Fun can't be added to something, like sugar to coffee or like songs to chores.

In their book *All Things Shining*, the philosophers Hubert Dreyfus and Sean Dorrance Kelly talk about this same issue in the context of

Wallace's depression and eventual suicide: in a secular age, meaning has to come from within, created from nothing.[16] Dreyfus and Kelly argue that this demand to make something wholly from nothing, by sheer force of will, is the weight that undid Wallace—and that the same demand is eating away slowly at all the rest of us, too. Wallace's desperate attempt to make fun equal delight on the cruise he took for *Harper's* or to see literary and televisual irony as a means to uncovering hypocrisy are the desperate grasps of the secular ironoiac who knows that dissatisfaction is only *your* fault, given the infinity of options available instead of ocean cruises and contemporary fiction and golf and air travel and all the rest.

But near the end of *Pale King* Wallace puts his finger on an alternative: boredom. Working through boredom in order to find deeper structure can form the boundary and contents for a playground. "It is the key to modern life," he writes, "If you are immune to boredom, there is literally nothing you cannot accomplish."[17] Embracing boredom doesn't really mean enjoying being bored. Rather, if we can bore through the boredom when paying attention to ordinary, mundane things (tax returns and televised golf are Wallace's examples), then something magical happens underneath. Ride that out, writes Wallace, "and it's like stepping from black and white into color. Like water after days in the desert. Constant bliss in every atom." Achieving this bliss requires giving yourself over to the structure of a situation rather than asking it to return its spoils to you.

This tactic is the cure for ironoia, too: you accept that meaning can come from *outside* of you rather than from within. Perhaps, even, that it *must*.

THE SUSPENSION OF FOLLY

Along with Mary Poppins, we assume that finding the fun starts with *us* rather than the thing itself. That we must bring something to the table that makes intolerable things tolerable, like a song covers chores or like chocolate covers broccoli. But as Suits suggests, a game is something good enough on its own, something for which on-its-ownness is precisely the point. Suits calls this willingness to accept the arbitrariness of a game the "lusory attitude." Golf would be worse than "a good walk ruined" were it a Broadway song and dance number about dropping balls in holes, Mary Poppins style.

To reject these systems as insufficient is certainly possible—the ironoiac does so all the time, after all. To doubt that anything positive, credible, or long-lasting could arise from something like the thankless, arbitrary world, or else to capture its surface effects and to see them as mere signs, exchangeable for one another via new proverbial blister packs of ironic appropriation. Or else to assume that you yourself are unprepared for them, not good enough for Pebble Beach—or anything else. It turns out we can hold *ourselves* at an ironic distance, too. But to issue such rejections just because the materials in our midst are annoying or unsatisfactory is to miss the point: they are *meant* to be annoying and unsatisfactory. Meaningful things contain something abhorrent, something revolting even though sublime. And every now and then, when you stare down their abhorrence, they will reward you.

In the summer of 2010, John Isner and Nicolas Mahut played a match of tennis at Wimbledon that lasted three days. *Three days*. Neither was able to break the other's service to tip the match out of equilibrium. The players served over one hundred aces each. Isner finally bested Mahut with a 70–68 final set, numbers more commonly found on a basketball scoreboard. It was an incredible moment, because Isner and Mahut had discovered something in Grand Slam tennis that nobody had found before, something preserved and durable, even as it was fragile and improbable, like a fossil at Pompeii. Two well-matched, world-class players with equally uneven volley games could make tennis go on almost forever. They coaxed the sport to give up this secret because they treated it with such ridiculous, absurd respect that the game couldn't help but release this meaning for all of us.

Contrary to what our intuition and our habits tell us, fun isn't accessed through facility —by choosing to do exactly what we want or by taking the easy path instead of the difficult one. In fact, the deliberateness and respect that produce fun result from deep dives into subjects rather than superficial explorations of them. Wimbledon 2010 taught us that lesson: persistence, repetition, and solemn attention allowed Isner and Mahut and all of us watching them to find the secrets hidden in the everyday. And experiencing those secrets is not a bad definition of fun.

But you don't need to be a professional tennis player. *Anyone* can treat *anything* with the deliberate attention that produces fun.

You already know something about my penchant for lawns. Installing landscaping is a one-time affair, but lawn care is habitual. And so, on the weekends, I mow the lawn. It's a chore, but doing it myself spares me the cost of a gardener and gets me outside and active. The prospect of owning a gas-powered lawnmower didn't appeal to me, partly because I don't want to cart around fuel for it, and partly because the cost, complexity, and mess of maintaining a combustion engine in a lawn appliance turns me off. So, I acquired a manual reel mower to reduce noise and air pollution, and to connect myself physically to the act of mowing.

It proved harder than my Dagwood cartoon–inspired memories of the practice led me to believe. The manual blades catch and stop short on my uneven plot, so that every few steps the whole apparatus jerks to a halt. At first I feared I'd made a terrible mistake. But over many Sundays, I discovered an optimal pace that allows me to keep the blades' momentum through the switchbacks.

An unexpected wrinkle arises—despite regular mowing and watering, the lawn isn't thriving (this lawn is a different one, by the way, of fescue rather than zoysia). An expert evaluates and concludes that the thatch thrown off by the mower is prohibiting water and nutrients from penetrating the soil. His prescription: bag and dispose of the thatch, particularly through the spring and summer seasons, which require more frequent mowing.

Easy enough, until I discover that the thatch catcher manufactured for my hipster mower alters the flow of my routine. Clippings get caught between the blades and the bag, or else they get strewn so far away from it that too much thatch stays on the ground, despite the new apparatus. Through experimentation, I find a slower cadence that produces better thatch throw-off, while improving the evenness of momentum I'd previously thought I'd nailed.

Then, later, after Mark advises me on the other lawn, I realize that the thatch routine was nonsense, and that the prior expert's expertise was fraudulent. Thatch is totally fine for the lawn, in fact: it adds nitrogen to the soil as it breaks down, nourishing it. The trick is not to leave too much thatch, a feat accomplished by mowing often enough that the new growth removed is modest (mowing often is better for the grass blades anyway), and to make sure that it doesn't suffer any defects, like fungus, which could spread via thatch. And so, back to square one, starting over with a new routine without the thatch catcher.

As I become one with my equipment, I find new goals and new accomplishments, refining the straightness of my lawn stripes over the months and the seasons. Even without the delight of neighborly jealousy, that great rationale for American lawn care, each week's mow offers the delight of a novel rendition of a familiar expertise.

It may seem ridiculous to call an activity like this "fun." But isn't it infinitely more ridiculous to imagine singing a Mary Poppins song to distract you from the drudgery of yard work? Just as golf is hardly a good walk spoiled, so lawn mowing (and mall visiting, and coffee brewing) are more than chores or refreshments. They are activities in their own right, the deliberate pursuit of which offers us access to their previously unseen depths.

Fun comes from the attention and care you bring to something, even stupid, seemingly boring activities. It's a foolish attention, even. An infatuation. Feeling like you are having fun is a sign that you have given something respect and that its creators have respected you enough to trust that you might be able to respect it. We fail to have fun because we don't take things seriously *enough*, not because we take them so seriously that we'd have to cut their bitter taste with sugar. Fun is not a feeling so much as an exhaust produced when an operator can treat something with dignity.

∎∎∎

In fiction, we often adopt the poet Samuel Taylor Coleridge's idea of a *suspension of disbelief*, a willingness on the part of a reader or a viewer to accept the improbable aspects of a narrative given enough of a "semblance of truth" to justify that suspension. Fun demands something more difficult. We know that novels and films and the like are invented, so we enter them already expecting to apply a generous grain of salt. But games, jobs, sports, and all the other experiences subject to the joy of fun are not fictional, at least not in the same way that stories are. Golf is an activity in which you *really do* strike a tiny ball with a stick on a giant lawn. Tennis is a sport in which you *really do* volley a felt sphere over a net on grass or clay or pavement. Real children *really do* have to clean their rooms. Real workers *really do* have to work their jobs, day in and day out.

Playgrounds offer a solution to this problem, because the boundaries they erect *are* fictional. Their contents are real, just as the contents of literature—characters and plot and places and language—are real. But the frame of a playground helps make the real world partly fictional. It wraps an imagined boundary around something, allowing us to suspend temporarily our ordinary relationship with it as ground, and to see it new, as figure.

Fun doesn't require a suspension of disbelief so much as a suspension of folly. So long as we are unwilling or unable to consider a set of actions as serious and intentional, even when those actions are mustered in the service of a seemingly absurd, foolish activity or end, then we will never be able to experience fun. As Mahut and Isner's Wimbledon match shows us, suspending folly is the first step in treating an activity with dignity, in taking it for what it is rather than what it isn't, and in following that conceit to its logical extreme. *Respect* is a more ordinary name for the suspension of folly. Fun is a feeling of respect. It is a feeling directed outward rather than inward. Likewise, when something doesn't feel fun, that feeling might really be a sign of our own premature contempt at an object rather than its failure to gratify us.

I know I'm guilty of this contempt. Aren't you? We expect things to come to us. To give us what we expect while also showing us something totally new. "Prove it to me," we say. "Prove that you're worth my time. . . . Sorry, too late. I've got better things to do."

Granted, some things don't even come close to meeting us halfway. We are used to allowing film, literature, art, and media to be "difficult" at times, to challenge our preconceptions. But ordinary things also resist us, and they do so more often—for example, suburban commutes, home decorating, sales presentations, and yard work. Just as the best art is difficult rather than easy, maybe all the other things that don't meet us halfway, or even partway, are *more* fun than the ones that give up all their secrets immediately.

...

So if we return to our newfound foe Mary Poppins, it turns out that the thing that makes the job fun is not finding the element of fun that makes it a game, but in finding the element of fun that makes it a job. Jobs are fun when they are not games or pastimes. When they are what they are, and when we take them seriously as such.

There's a truth in tautology. The job's a job. The commute's a commute. Really and truly taking something as a unique and discrete thing in the world that is nothing other than what it is. We tend not to have fun because we fail to take things seriously. And (yes, ironically) the failure to take things seriously is also the source of ironoia. If irony represents the crack in the universe through which distrust and anxiety about living in a world full of surplus arises, then fun offers a glue with which we can seal those cracks and restore dignity to all the things we encounter—including ourselves.

Remember the fool? The fool may seem like an imbecile or a stooge, but the true fool is infatuated: "I'm a fool for you," as we say. This obsession is affectionate and earnest rather than optimistic or naive. The fool's fun is also related to fondness: it requires devotion and enthusiasm. Fun is a commitment to something accidental. It demands seeking out novelty within the boundaries of playgrounds. Fun is the cold, indifferent stupidity of something that is just what it is. Fun is exploration. Fun is finding tiny air bubbles of freshness in a suffocating familiarity. Fun is treating things like their existence is reasonable. Fun isn't extracted, like sugar drained from cane, but enacted, via the process of harnessing the oscillating temptations of irony's earnestness-or-contempt, and swapping them out for the commitment to discover novelty in something familiar.

Fun can only arise when we treat the things we do or make as *exactly what they are* rather than trying to hide it. That's why fun isn't a kind of pleasure—not a direct one anyway. Fun is coming to terms with something unreasonable and arbitrary. It is giving respect to something that doesn't deserve it or becoming infatuated with something for which infatuation seems impossible. By attending to it. By working it carefully and deliberately, over time, in the hopes that it might blush before you and release its secrets.

FOUR

Play Is in Things, Not in You

Play is possible when freedom is limited rather than expanded.
It is not the opposite of work, nor the opposite of depression.
Play is deliberately working with the materials we encounter.

SIMPLE, ACCIDENTAL EXPERIENCES LIKE MY DAUGHTER'S EN-
counter with the mall underscore a profound truth about fun,
something that offers us a way to rescue Mary Poppins's "snap" and
expand it from magic into method. As we already learned, "finding
the fun" in an activity like mowing the lawn or navigating the mall
is a process of treating the situation as exactly what it is rather than
attempting to reject it in favor of a more distracting or entertaining
alternative.

You might stop me and object: wasn't my daughter just avoiding
the drudgery of mall transit by finding (or inventing) a substitute?
While we were supposed to be progressing through the shopping
center, she was instead pursuing an orthogonal end. Parents are

accustomed to this sort of sideshow, and we usually frame it as a distraction. "Stop dawdling!"

But in truth, my daughter's playful act interpreted her situation as much as my attempt to rush us through it did. While I was only focused on the goal, she managed to attend to the process. And her experience of this apparently meaningless act was clearly the more fun approach between the two. Whereas I rejected the scenario and attempted to seek escape from it, she stared down her boredom and embraced what emerged beyond it. The freedom of that particular playground came from her willingness to submit to the mall and its materials, rather than from her ingenuity in devising a way to overcome the boredom she found there. When we play, we give ourselves too much credit. Play is the opposite of irony: rather than distancing ourselves from things, in play we draw them close and meld with them. We give ourselves over to them, even, subordinating our own agency to a larger system.

It sounds miserable, but it turns out to be fun. Or more precisely: we do not arrive at fun by expanding our circumstances to allow for less wretched alternatives, but by embracing the wretchedness of the circumstances themselves. We might even go so far as to say that fun comes from wretchedness. Without a starting situation that resists us, it is impossible to produce meaningful, let alone enjoyable, experiences. Fun *is* impoverishment, blight, and squalor. Fun is the broccoli *without* the chocolate, once we realize that the broccoli itself is not an end but a resource we can put to use.

Somehow we got the idea that fun refers to enjoyment, and that games offer us special access to that enjoyment. But games and fun are not connected because games are intrinsically entertaining or enjoyable, but because games are already intimately associated with playgrounds. Games are experiences we encounter through *play*.

Play is the act of manipulating something that doesn't dictate all of its capacities in advance, but that *limits* its capacities through focus and exclusion. Minecraft asks you to survive in a world made of inhospitable cubes you can use as resources. Candy Crush asks you to solve a puzzle given a limited supply of powers. Golf asks you to get a tiny ball into a slightly larger hole many hundreds of yards away by striking it with a stick.

Despite its intrinsic connection to games, play isn't limited to them. Play is everywhere, in anything we can operate—espresso machines, lawnmowers, shopping malls, anything. It is not an act of diversion, but a name for the feeling of making something work, of the results produced from interacting with its materials. That's why we also say that we "play" an instrument or a sport, why we rejoin our children for "playing" with their food, or even why we can talk about sexual self-stimulation as "playing with yourself."

...

As IT TURNS out, this more general notion of play has a long history. The twentieth-century philosopher Jacques Derrida used the term *freeplay* to describe the possible perspectives toward cultural works: "by orienting and organizing the coherence of the system, the center of a structure permits the freeplay of its elements inside the total form."[1] Derrida was mostly interested in the way the manipulation and interpretation of symbolic systems produces meaning (like many of his contemporaries, he called any such system a "text," whether or not it involved words and language).

His point is that meaning exists only in reference to its contexts. A poem or a painting has specific words or images or forms or pigments, which "center" the structure and form its materials. But the meaning of these "systems," as he calls them, exists inside another structure, such as the specific community or cultural context in which it is experienced. For example, in 1854, Henry David Thoreau's book *Walden* endorsed simple living for the sake of self-sufficiency. A century and a

half later, *Walden* has become something of a bible to ecological preservationists in the era of climate change. The modern reader can't
help but think of it as a collectivist credo rather than an individualist
one.

Derrida's philosophical system, called deconstruction, gets a bad
rap for what some perceive to be a nihilistic, anything-goes attitude
about meaning. To its critics, deconstruction is an excuse to make
up any story you want about a work of art, and this perceived ability to make-believe is why the philosophical trends of the twentieth
century known as postmodernism and poststructuralism are often
dismissed as lewd versions of extreme relativism.

But put aside the criticism you've heard regarding these names
and terms. Derrida teaches a lesson worth learning: a multitude of
latent meanings subsist in "texts," but *credible* meanings must relate
to the text, coupling new observations to it so that those interpreta-
tions work like the mechanisms in a machine. A book like *Walden*, for
example, can be read as a transcendentalist, American Romantic text
circa 1860 or as an ecological text today. For Derrida, play is a name
for a text's ability to bear new meanings for different readers in different contexts. The play theorist Brian Sutton-Smith discusses a similar
phenomenon, the "play of light, the play of the waves, the play of
components in a bearing case, the inner play of limbs, the play of
forces, the play of gnats, even a play on words."[2] Sutton-Smith relates
this use to the original meaning of the German word for "play," *Spiel*,
as "dance."

The game designers Katie Salen Tekinbaş and Eric Zimmerman
adopt this sense of play too, in their formal definition of the concept,
which is really the same as Derrida's, but without all the philosophical
window dressing: "free movement within a more rigid structure."[3]
When designing a game, the point is not to make it taste sweet, but
to fashion a structure. And when playing a game, the question is not
how to overcome that structure, but how to subject oneself to it—

like a golfer at Pebble Beach, like Isner and Mahut at Wimbledon, like my daughter at the mall.

Machines and apparatuses are good metaphors for understanding play. When you operate a mechanism like a steering wheel, the device has some "play" built in: a space through which the steering wheel can be turned before the steering shaft couples with and turns the pinion at its end. Likewise, the play in a guitar comes from the manipulation of a set of differently gauged strings held under tension across a fretboard. We certainly manipulate our cars and our guitars, our bodies and our language. But before we can do so, we need some machine or apparatus upon which we can exert force. A playground in which to play.

And when it comes to play, we give ourselves too much credit: the play is in the thing, not in us.

POLICING PLAY

This idea—that play is a property of things rather than an experience had with them—runs against everything we've been taught to believe about play. Play is thought to be central to the development of children, to the point that a scare has been brewing over the decline of play in contemporary culture. Writing in *Aeon Magazine*, the psychologist Peter Gray laments the loss of the "hunter-gatherer" education he experienced as a child in the 1950s.[4] The freedom to explore the outdoors, to discover new ways to make use of time to stave off boredom, and to pursue one's own chosen interests and media—for Gray, all these activities have been dampened or eliminated, pushed aside by increased structure and oversight in school; highly organized and usually adult-directed activities like sports, Scouts, ballet, and music; and the slow, methodical reduction of the expanse of physical environments to which kids have unfettered access.

As competition for test scores and college entry requirements have swelled, parents, educators, and governments alike have increased the structure of children's routines, so that the average middle-class kid is scheduled from dawn to dusk and beyond with school, homework, tutoring, and extracurricular activities. The changing global economy has reduced options for viable middle-class careers, making college the only option for upward-mobility (or even just mere stability). Meanwhile, long-term reductions in public funding (in the UK and Europe as much as in the United States) have made college increasingly expensive for ordinary folk, which only further accelerates the fear that kids will miss out on already dwindling future opportunities unless they invest even more time in advanced courses, test preparation, and organized, official-looking extracurricular activities.

Even if kids had time left after all of this regulation, they're increasingly unable to partake in the world outside their schools and homes anyway. The 1950s hunter-gatherer childhood of Gray's memory is partly a nostalgic myth in the spirit of Thoreau, for one part. For another, kids are largely prohibited from meandering on their own or in groups today. Writing in the *Daily Mail*, David Derbyshire contrasts a contemporary eight-year-old schoolboy (Edward), with his great-grandfather (George) of the same age.[5] In 1926, George was able to meander some six miles to a pond to fish. Eighty years later, Edward is driven everywhere, even to safe, predetermined venues for bike riding.

This shift didn't happen all at once. Edward's grandfather Jack was afforded a mile of freedom from his house at age eight, in the 1950s. His mother, Vicky, was allowed to wander about a half-mile away, to the local pool, in the late 1970s. By 2007, little Ed was permitted to stray less than three hundred yards from his door, as far as the end of the street.

Parental choice—and fear—is to blame, in part. But restrictions like those that sequester Edward have also found their way into previously flexible environments, like schools. Thanks to the pressures of high-stakes standardized testing, overcrowding resulting from reduced tax bases, and reductions in teacher and staff support, some schools have felt forced to institute prison-like regimes. Recess has been reduced or eliminated. Lunch periods have been shortened to allow for multiple sorties of kids to use facilities designed for far fewer students. To maintain timeliness in these new lockups, some schools have mandated silent lunches, because kids eat faster when they are not allowed to socialize.

Not everyone thinks holding kids hostage is good for them, but attempts to return even to the historically modest standards of Edward's mother's era pose new challenges. In recent years, attempts to resist the limitations imposed on kids have met with terrifying social and legal consequences. Maryland parents Danielle and Alexander Meitiv describe themselves as proponents of "free-range parenting," a philosophy of childhood independence that hardly would have needed formalization in great-grandpa George's day. Among their newly radical acts, in April 2015 the Meitivs allowed their six- and ten-year-old children to walk home from a local playground alone. The police took the children into custody, and Child Protective Services investigated the parents for possible child neglect.[6] It was the family's second run-in with the authorities on the matter in less than six months; the first time, they were "found responsible for unsubstantiated child neglect"—for allowing their kids to walk home from the park.

The Meitivs became paradigmatic of overboard reactions to even modest freedoms for kids. But they're hardly alone: thanks to the ubiquity of mobile phones with which to call in "concerns," mixed with our tendency to be unacquainted with our neighbors, everyone is afraid before they are generous. Irony is a way of life, even for grandmas in suburbia. Combine this anxiety with decades of classroom and television marketing about "stranger danger" (despite a substantial drop in an already very low likelihood of abduction), and

Gray's hunter-gatherer childhood is being policed into oblivion. And not without consequence. Gray correlates reduced opportunities for play over the last half century with rising rates of childhood mental disorders, including depression, anxiety, and suicide, which have increased four- to eightfold since the 1950s.

While a majority of research connects the need for play and exploration to childhood development, other studies have pointed to the need for play among healthy adults as well. Ordinarily, we oppose play and work, committing a fallacy that Stuart Brown, head of the National Institute for Play, calls the "work-play differential."[7] We think of play as synonymous with leisure, distraction, and waste. And given that we live in a society obsessed with improving efficiency and productivity, it's no wonder that play, games, fun, and their brethren terrify our culture at its foundation. In a line less quoted but just as quotable as Mary Poppins's spoonful of sugar, Brian Sutton-Smith enjoins us to believe that the opposite of play is not work, but depression.[8] It's a sentiment adults would probably like their bosses to believe, as we all try to organize any viable escape from our overtaxed lives in offices and in front of computers.

Brown, a medical doctor with experience studying murderers, argues that a deprivation in play contributes to mental anguish, which in turn leads to homicidal thoughts and actions. As is so popular these days, Brown relies on cognitive explanations to justify the introduction of "play hygiene" to help cure us of our play-deficient lives. Using the evolutionary experience of play in the development of animals—the way creatures test out social behavior through play—Brown argues that play "lights up" the brain both at its animalic base and in its human frontal cortex. We err particularly, according to Brown, when we relegate play exclusively to the purview of children. As Sutton-Smith's quip-quoters hope to encourage and as Brown hopes to prove through cognitive science, a dearth of play might not only be annoying or disappointing, but also tragic and even dangerous.

In an age of efficiency, it's easy to connect the decline of play to the rise of despair. Today, we bring our work home on our smartphones and laptops so that the two lives are increasingly indistinguishable. Even at the office, our actions and activities are monitored and instrumented, whether directly with hard controls like quotas or targets, or through softer ones like open-plan offices, in which everyone's activities are visible to everyone else. In schools, an almost fanatical obsession with measurement through standardized tests has endorsed new regimens to reduce recesses and other unstructured activities. Teachers are transformed into the same kind of monitored workers that fill cubicles, their performance assessed from afar, out of context. Meanwhile, a paralyzing fear of the very violence and unrest Brown suggests might arise *because* of the absence of play has only further endorsed the treatment of students and workers as inmates.

A SUPPOSED SALVATION

Johan Huizinga, the early twentieth-century Dutch anthropologist whose endorsement of fun we already encountered, situates play at the very center of human activity. Huizinga sees play as the means through which culture *itself* is produced rather than an activity of distraction we pursue when we grow weary of work. Man is not primarily a knower (*Homo sapiens*) nor a creator (*Homo faber*), but a player—*Homo ludens*. For Huizinga, play is a free activity separated from ordinary life and done for its own sake. But despite this freedom and separation, play creates order—indeed, play relies on structure in order to produce that order.[9]

Huizinga shows that the rituals and practices of human culture, from law to religion to war to politics, all rely on the elements of play as fundamentals. In law, for example, the roles of judge or prosecutor involve the adoption of specific garments and accessories not unlike those of the court jester or the theater actor—another domain we describe with the word "play." The carriage of justice in a trial bears much in common with theater, in fact—it has all the trappings of scripted entertainment, which is part of why we find courtroom television dramas so enjoyable. Legal trials involve battles of wits and verbal performance, they subject the accused and the accuser to the accident of chance, and they take the form of a contest through which a winner and a loser is definitively and unequivocally determined.

If Huizinga is right, then play advocates like Gray and Brown, along with practitioners like great-grandfather George and the Meitivs and their children, are onto something. "When we play," the play scholar Miguel Sicart summarizes, we "appropriate the world, we make it ours, we express ourselves in it, we make it personal."[10] We need play, Sicart advocates, in order to help us take a distance from life and see it anew. We already embrace play in circumstances of grave consequence, like courtrooms and temples. Why not in neighborhoods and offices too?

But unwittingly, Sicart also puts his finger on the problems with much of the contemporary advocacy for play. Play is undertaken for the sake of the individual or the group or the society. And the playground itself—the forest, the swing set, the conversation, or whatever—always serves as a means to that end. The story of great-grandfather George offers an instructive example. In Ed's and Vicky's and Jack's and George's cases, the objects of supposed freedom—a fishing pond, a woods, a swimming pool, or a street—are invoked as mechanisms for roaming and its concomitant psychological benefits, the ones Peter Gray and Stuart Brown endorse. The material world, it would seem, is just a pastoral big-box store. A venue for securing materials for incorporation and consumption.

But Huizinga sees play as an experience separate and distinct from ordinary life, one pursued absent material gain. Play in this sense is *autotelic*, a term for activities pursued for their own ends, rather than as mere means. (Csikszentmihalyi's flow is often also called autotelic; high-performance, flow states, the theory goes, are easier to achieve for people predisposed toward self-regulated goal setting and pursuit.) It's tempting to understand play as an activity that tousles the hair of presence-at-hand, rejecting the instrumental, purposeful uses of materials like shopping malls and roads to and from the park or the fishing pond. But that's not entirely the case. One need not look any further than Gray's or Brown's endorsements of play to see that the practice is conducted "for its own sake" only in part. The benefits of play are expressed mostly in terms of their resultant underwriting of mental (and often, physical) healthfulness.

It's impossible to fully separate ourselves from the things that surround us. Indeed, presence is not a more desirable philosophical orientation for Heidegger (and Derrida, whose philosophy is built atop Heidegger's), but an error. It is dangerous to demand that everything always be useful, and that its utility be discerned and predicted in advance.

If we advocate for play as such, as do Gray and Brown and Sicart and others, we risk replacing the autotelic experience of play in all its individual diversity with a more instrumental, generic version of play. Play becomes a skill or literacy, akin to critical thinking or problem solving. In addition, play becomes politicized, turned into an ideological talking point meant to oppose the rigid, inflexible regimens of contemporary policy and law. As the Meitivs and the public school silent lunches attest, contemporary law and policy have plenty worth opposing. But too often, play itself becomes a kind of trump card cast against overzealous regulation, so that it risks becoming its own ideology.

Most commonly, play is mustered as a palliative to structure, the supposedly free and open and self-directed nature of play restoring an apparently lost creativity that Edward's generation needs to regain from George's. Gray cites an extreme example, the Sudbury Valley

School in Massachusetts, at which students from ages four to nineteen are let loose all day, every day, to do whatever they want instead of following coursework, preparing for tests, or queuing up for silent lunches. Here's Gray selling the idea, and convincingly, too:

> To most people, this sounds crazy. How can they learn anything? Yet the school has been in existence for forty-five years now and has many hundreds of graduates, who are doing just fine in the real world, not because their school taught them anything, but because it allowed them to learn whatever they wanted.[11]

Sudbury Valley is a private school that charges roughly $8,500 per student. It's not as expensive as the most elite schools, but it does suggest some self-selection among its ranks: wealth is the best predictor of educational success. Notwithstanding the possibly hypocritical fact that the school embraces at least enough formal structure to maintain a ten-acre campus and to charge tuition, the Sudbury model of student-directed learning and governance shares less in common with traditional schooling and more in common with *unschooling*, an educational philosophy first advanced in the 1970s by the WWII naval veteran turned grade-school teacher John Holt. Unschoolers hold that children possess an innate curiosity and drive, which institutions like formal schools only beat out of them. Thanks to its inherently unstructured nature, unschoolers are often grouped together with homeschoolers in the imagination of the public, even though many homeschoolers maintain an educational regimen as structured as today's public schools.

As in any community, there are successful unschoolers. Among them is Jedediah Purdy, whose 1999 book *For Common Things: Irony, Trust and Commitment in America Today* offers a paean to sincerity so mushy it would make even Christy Wampole leer toward contempt.

In Marshall Sella's profile of (an admittedly still very young) Purdy for the *New York Times Magazine* in 1999, Purdy came across as the Thoreau-endorsable, American woodlands apotheosis of Pasolini's rhetorical boy whose things make him who he is. Raised in rural West Virginia, "there is no vine that Purdy cannot identify," wrote Sella.[12] Purdy, born in 1974, was just the right age to take part in the Romantic, self-directed learning that his parents encouraged, sheltered from popular culture and immersed in the purportedly more real reality of woodsheds and hog slaughters.

But how does Purdy's story end? At thirteen he tried out public school, where his lack of common experiences and cultural references alienated him from his peers. So he attended Exeter, then Harvard, then Yale Law. He is now the Robinson O. Everett Professor of Law at Duke University. Which is not to say that the West Virginia woods are incompatible with Harvard Yard. But if the embrace of play and freedom ends up euphemizing the pursuit of entrenched, monied, polite society, then we should avoid the circumlocution. The best way to eschew irony, perhaps, is to have the luxury to do so.

Still, Gray's hope for Sudbury and Purdy's early start as an unschooler hold some truth. A situation so confined and barren that nothing can be manipulated within it is not a playground but a prison. And many of our schools and workplaces do sometimes feel like prisons. But in our rush to reject that imprisonment, we might be overcorrecting. As golf courses and shopping malls and yards and coffee shops all show, playgrounds aren't places where you can do anything whatsoever. Pure, abstract play is a fantasy. After all, Gray's endorsement of play in school environments mostly suggests that various levels of structure and organization are available for rearing and educating kids, and that alternatives to total imprisonment might have some merit—hardly a eureka moment for anyone who has discovered Montessori or backyards or 4-H or Lego.

PLAY IS SUBMISSION, NOT LIBERATION

Why insist that we choose between Walden and Walmart? Perhaps our highly regimented lives have made free play seem like an escape hatch from suffocating oppression. And, ironists that we are, we have mated structure (which is considered restrictive and therefore bad) to institutions like schools and businesses, while dreaming that the free-form domain of nature and whim is liberating. Such a move reinforces the work-play differential, treating play as a release valve for labor. But worse: it suggests that no real salvation is ever possible for institutional practices like work and school and errands and yard care and all the other activities that fill most of our waking time.

Adults' jealousy for children's play arises from the mistaken assumption that children "do what they want" in the first place. Play advocates make this same mistake when assuming that the practice is de facto subversive. "Play matters," Miguel Sicart begins one such exhortation, "when it is appropriative, taking over a situation and turning it into a context for play."[13] Toys, Sicart suggests, begin developing this practice. The ball or Frisbee or bicycle "modify the space in which they are used for play."[14] And it's true: just as my daughter's mall game modifies the shopping center into a temporary arena, so the Frisbee can turn the backyard into a pitch and the bicycle can turn the trail into a race course.

But are such maneuvers really appropriative? Just as Gray and Brown suggest that play must rescue childhood from the oppression of institutions, Sicart implies that toys issue tiny rebellions, seizing or annexing the otherwise controlled, orchestrated universe and rescuing it for the player, whose heroism finally allows oxygen into the heretofore suffocating tyranny of backyards and roadways. Make sure your neighbors get the memo that their grills and driveways are mere instruments of a spectral, pervasive fascism.

...

THE GAME DESIGNER and critic Mary Flanagan holds a similar but even more extreme version of this position. Subversive play, she suggests, invites or requires one to play against the intention and authority of the system. Flanagan suggests the term *unplaying*—a notion that mirrors unschooling in more than name—for the practice of taking a game or a toy and undermining its intended purposes with alternative ones.[15] While Sicart's aims are more contemplative, Flanagan embraces a clear directive to unplay in order to subvert the purportedly mainstream (and therefore ostensibly oppressive) properties of traditional games.

Flanagan connects unplaying and subversive play with twentieth-century avant-garde art, which embraced similar political aspirations. In particular, the movement of the 1960s known as Fluxus offers inspiration, partly because it too used games as objects for subversive commentary. Reimagined chess sets were a common trope of Fluxus, for example. The Japanese artist Takako Saito's sets replace the traditional pieces with unexpected alternatives, such as spice jars or vials of liquor. Instead of recognizing the pieces by their conventional shapes, players must smell or taste them as the game goes on to distinguish them from one another.

But the most famous Fluxus chess set is probably Yoko Ono's 1966 Play It By Trust, also known as White Chess Set. Both players' pieces are white, as are all the board's squares. Such a design was meant to make the two players, who normally symbolize military opponents, indistinguishable. The work has been widely interpreted as an antiwar statement, because the conceit of opposition is undermined once the pieces begin to consort on the field. On the battlefield of humanity, the game implies, violence and opposition are pointless, invented affairs.

But far from undermining chess, Saito's and Ono's chess variants mostly serve to demonstrate the flexibility and resilience of the original game. Saito's Liquor Chess isn't only an exercise in memorization; it also juxtaposes chess as a test of wits with drunkenness, the easiest and most culturally acceptable way to deliberately undermine mental acuity. You can purchase many varieties of liquor chess at novelty stores and museum shops—the Exeter set's polite version of beer pong.

As for White Chess Set, it works well as an antiwar sentiment under the critic's pen or enshrined atop a gallery plinth where it is meant to be looked at rather than played. But in practice, the game offers as much of a depiction of combat as it does of pacifism, thanks to the interpretive play of meanings we learned about from Derrida. White Chess could be seen as a depiction of the fog of war in conventional warfare, for example, or the inability to discern enemy combatants from friendlies or civilians in guerilla warfare and non-state-sponsored insurgencies. Or players could abandon the conceit of war entirely and enjoy White Chess as a Machiavellian logic puzzle, a metagame for traditional chess, in which the player must persuade or deceive an opponent to acquire control of the most desirable pieces at a given moment.

Materials are harder to undermine and subvert than skin-deep opposition might suggest. True, it's often appealing to push the boundaries of something, to test its limits. The foolhardy youth who drives too fast to feel the sensation of risk and to give the finger to authority, despite the danger; or the marathon runner who pushes her body to the edge of its abilities in order to increase her future capacity; or even the artist—like Ono or Saito, Rauschenberg or Manzoni—who pushes the limits of recognizable patterns for painting and sculpture in order to make the institutions of art do new tricks.

But these maneuvers ironize more than they subvert. Or better, to think of them as subversive is to contain and sterilize them under the plastic-wrap of commentary or sabotage. White Chess Set undermines neither chess nor war; mostly it succeeds as a static sculpture created by a famous artist. Avant-garde art was the first bastion of irony, and it continues to breed ironoia.

Likewise, Sicart is wrong about toys like Frisbees and bicycles. To think that toys reconfigure the world around us presumes that that world is somehow already fixed and settled, for one part, and unworthy of our attention, for another. While it might seem like a moral high ground to lament using roads for automobiles rather than bicycles, and lawns for decoration rather than Frisbee, and malls for improvised dancing rather than commerce, reality proves harder to pin down. Just as Walmart proves that the nobility of things might be found in our willingness to commune with them rather than in their intrinsic properties, so toys and art and other supposed disruptions show us an approach to identifying and embracing the worldful attitude more frequently and with greater delight.

THE MAGIC CIRCLE

Johan Huizinga offers a useful, general concept for this approach to physical therapy. His idea is an anthropologist's technical term for a playground. In his book *Homo Ludens*, Huizinga has this to say about the nature of play:

> All play moves and has its being within a play-ground marked off beforehand either materially or ideally, deliberately or as a matter of course. Just as there is no formal difference between play and ritual, so the "consecrated spot" cannot be formally distinguished from the playground. The arena, the card-table, the magic circle, the temple, the stage, the screen, the tennis court, the court of justice, etc., are all in form and function play-grounds, i.e., forbidden spots, isolated, hedged round, hallowed, within which special rules obtain. All are temporary worlds within the ordinary world, dedicated to the performance of an act apart.[16]

By "play-ground," Huizinga doesn't mean the sand-encircled space in the park down the street with the slides and swing sets. This modern notion of the playground as a park-like environment with physical equipment and open spaces for kids didn't exist until the nineteenth century, when playgrounds first appeared in Europe and America.[17] These public- or school-accessible areas were erected for both moral and safety reasons. For one part, the playground was a place where rules, manners, and sportsmanship could be exercised and policed. For another part, public streets, where children had previously played, became feared for breeding hooliganism and grave physical danger, especially after the introduction of the motorcar. Huizinga uses the term much more generally, in a way that applies to any historical period. He's suggesting that anything whatsoever can be construed as a ground for play, where play becomes possible once the materials at hand are taken seriously and manipulated with deliberateness.

In the middle of this passage Huizinga lists a series of examples of these sorts of deliberately marked-off playgrounds, many of which illustrate his larger argument about play's central role in the development of human culture: "The arena, the card-table, the magic circle, the temple, the stage, the screen, the tennis court, the court of justice, etc." Most of these are quite specific, naming particular locations where the stakes of culture play out, so to speak. Gladiatorial combat and sport take place in the arena and the tennis court; gambling at the card table; ritual and religion in the temple; mimicry, representation, escape, and carnival on the stage and screen; justice and corruption, absolution and incarceration in the court.

■ ■ ■

BUT ONE OF these playgrounds is not like the other, and we've encountered it before: the *magic circle*. It too has a specific cultural meaning: it is the ritual space marked out by magicians and sorcerers, such as those who practice Wicca. When used in sorcery, the magic circle has a relation to other sacred traditions, including the Hindu and Buddhist mandala, which is sometimes used as a symbol of a sacred space or to mark out a locus for meditation. But as the most abstract and, arguably, the least familiar of Huizinga's examples of playgrounds, the magic circle has taken on special significance in game design.

Katie Salen Tekinbaş and Eric Zimmerman adopt the term to refer to the special place in space and time that is created when a game takes place.[18] That place could be the hopscotch or foursquare court, or it could mean embracing Suits's lusory attitude when reconfiguring a field as a pitch for a friendly game of soccer, or it could mean adopting formal rules and regulations, as in the case of a World Cup football match or a family game of Risk.

Among those who make, study, and play games, the "magic circle" has come to signify the boundary between the ordinary world and the game world, and the concept is often used as a tool for discussing the mechanics or ethics of playing a game. For example, does it "break" the magic circle if a player of Words With Friends, the popular smartphone variant of the board game Scrabble, looks up possible plays on an anagram-solving website? The magic circle has also been central to debates about the intermixing of real and virtual worlds. Players of a massive multiplayer online game (MMOG) like World of Warcraft can buy more advanced characters rather than earning them by rising in the ranks through the grind of repetitive progress. What is the relationship between playing the game and the ethics of endorsing or abiding the factory-like conditions in which Chinese workers "manufacture" such characters for sale on the open market?

But in deliberating the pliability and security of the magic circle, game designers have missed a more useful interpretation of Huizinga's magic circle: it names a generic process of circumscription.

The arena and the court and the theater are complex, permanent, and inflexible tools, which, once erected, service specific activities. To bring them about requires great effort. But the "marking off of a sacred space" need not involve landscaping and cement and wood and velvet. In addition to material construction, Huizinga allows for the marking off of a playground to be done "ideally"; that is, as conception rather than construction.

"Magic circle" is perhaps too dramatic a name for the more ordinary process of material or ideal circumscription. That's why I embrace *playground* as an alternative. The circumscription of play is really just the context of particular uses or states of things. When my daughter reconfigured the shopping mall to service her ends, she did so by taking a portion of the context around her and drawing a conceptual line around it. On the inside of that line—within the playground—an invisible, imaginary membrane held the tiles, her feet, the crowds, my momentum pulling her, and so on. Outside the circle, everything else still persists, including all the factors that intervene and interfere with her game, vying for her to adjust the shape and position of the membrane and incorporate them.

"When we are playing," Sicart insists, "anything can become a toy."[19] It's a tempting conclusion, because who wouldn't want life to be more playful, more free and liberating? Except the flipside of liberation is imprisonment. One person's liberation becomes another's anguish. The rush-hour drivers who must negotiate with the cyclists "liberating" a road in a city poorly designed and zoned for bicycling. The older couple hoping to use the park's perimeter for leisurely calisthenics who are unexpectedly sideswiped by a flying Frisbee. As we learned from Mary Poppins, turning the world into games (or toys) is a self-defeating goal. Instead, we should strive to encounter any circumstance as what it is rather than believing we can transform it into something it isn't.

...

IN THE KITCHEN just now, my wife gently rejoins our baby daughter, who sits in her high chair eating dinner. Individual, wrinkled green peas dance across her plastic tray as she bats it hard with each fist: *Bam! Bam! Bam!* The peas dutifully lurch up and to the floor, to the delight of my daughter. "No!" my wife rejoins futilely, as if the baby can understand. "If you're not going to eat the peas, then you're done with dinner." It turns out the baby is done with dinner.

It's one of an endless set of examples of a familiar pattern. "Don't play with your food." Food, after all, is for eating. Not throwing or sculpting or moving around on the plate. At least, such is the primary purpose of food, the purpose that justifies our growing, buying, preparing, and eating it. But then again, as every child and viewer of *Close Encounters of the Third Kind* knows, food is in fact *also* for throwing or sculpting or moving around on one's plate. Peas and mashed potatoes and chicken and applesauce invite us to explore their material properties as much as their nutritive ones. The fact that peas fall, that potatoes squish, and that condiments splatter are all features of foodstuffs that cannot be denied, even if they can be regulated via household or cafeteria policing.

Even adults play with their food, once play is understood as the deliberate exploration of something as a playground. We celebrate or lament the mouthfeel of an oyster or a Cabernet. Yes, true, the contexts for appropriate food play might differ for adults, for children, and for babies—and for dogs and birds and outdoor grills, for that matter. The sensation to reject or rejoin an act of physical therapy—"don't play with your food"—is also a reminder that circumscriptions other than the ones that we actively and willingly partake in are possible—and even desirable. They offer invitations to turn ground into new figure, even if just temporarily. Ironoia flares up when we fail to allow an object, event, or situation to be what it is. But play invites us to find new purchase on those situations.

My daughter's food throwing is a nuisance. It's not what food "is for," within the context of a playground that presumes food is for the feeding of a child. But it's also precisely what food is for, especially for a baby. Feeling the texture and viscosity and pliancy of foodstuffs in her hands. Inspecting the colors and shapes. Making contact between squash or blueberry or Barbara's Puffins Cereal and tongue and mouth and teeth, often for the first time. Just as Boniakowski's Cheetos become newly valid when subjected to the ironized transformation into molecular gastronomy, so a child's food play becomes tolerable once presented within the frame of haute cuisine tasting menus or the blather of fine-wine sommelierdom. But this perspective on food is just another circumscription, one in which working the material of foodstuffs—their feel, texture, shape, color, and arrangement—also welcomes other materials: the expertise of world-class culinary arts, to be sure, but also appeals to social signaling, conspicuous consumption, sophistication, seduction, and all the other exhausts of adult food and drink.

There are differences between my baby playing with her food at the kitchen table and me playing with mine at a Michelin-starred restaurant. But the primary difference is my own willingness to take the latter sort of play as serious and legitimate, and the former type as merely transitional and irritating, a thing to get over rather than to embrace. Ironically, the reverse could as easily be the case, given time. Fast-forward ten years hence, when I attempt to introduce my baby-turned-child to the delights and customs of fine dining—dressing up, using the various tools of a formal place setting, partaking in the various courses of a meal that lasts over hours rather than minutes, and so forth. Now it will be me begging her to play with her food in a different way, an invitation she will find as unfamiliar and unpalatable as I do her blueberry crushing and toast flinging.

EVERYTHING IS AT PLAY, ANYTHING CAN BE PLAYED

Peter Gray and Stuart Brown want to show that play is a necessary activity for regulating cultural contentment. Miguel Sicart wants to show how play can wrest meaning in a culture that is otherwise deadened or inactive, rendering it newly creative and alive. Mary Flanagan hopes to undermine repressive mainstream voices through subversive "unplaying" of specific targets. In each case, play takes on the role of liberation, freeing otherwise subjugated individuals into the comfort of creative self-actualization.

But Huizinga offers a far more mundane and therefore more powerful insight. He contends that play is the manner by which culture itself gets produced. Play is not an alternative to or a respite from work, but the process through which work is done—including the work of play in the sense of leisure and release. Play is not an activity opposed to work or productivity, as our intuition tells us. It is also not an automatically freeing, liberating activity that rejuvenates us from the work of ordinary life in order to return us back to it. Instead, play is a more general concept. It is the act of working a structure, rather than the act of working that structure for the purpose of leisure, distraction, rejuvenation, or even enjoyment.

Even if play produces fun, the basic experience of play is not letting loose or doing whatever you want, but carefully and deliberately working with the materials one finds in a situation. Play is not only fun, not only a child's activity, but also exploring the free movement present in a system of any kind, where *system* might refer to a social situation as much as a machine assembly. Play is the process of exploring such a system, whether it be a toy, a game, the form of the sonnet, or the economics of a household. Thinking worldfully, it is better to think of play as a condition of the universe rather than a human activity—everything is "at play."

And yet the work-play differential still proves a hard habit to kick. People who endorse play and fun as general concepts still hope to lift play from lower to higher cultural status. But in so doing, they rein play back into the corrals from which they hoped to rescue it. The game designer Raph Koster reinterprets *fun* as a type of cognition and learning: the good feeling we get when we identify patterns and solve problems. That we learn through play is a common sentiment.[20] "Play is our brain's favorite way of learning," writes the author Diane Ackerman in an aphorism emblazoned in large letters upon on a wall in the National Museum of Play in Rochester, New York. Even play proponents like Brown are guilty of recuperating play back into productive service. Just as soon as we utter suspicion of the work-play differential, play gets put back to work, rescuing us from work in order to make us happier and more productive.

∎∎∎

PLAY IS NOT an act of diversion, but the work of working a system, of interacting with the bits of logic within it. Fun is not the effect of enjoyment released by a system, but a nickname for the feeling of operating it, particularly of operating it in a new way, in a way that lets us discover something within it, or to rediscover something we've found before. This latter experience is hard to explain, like lightning striking twice; the improbable experience of finding the same permutation in a subsequent run of a complex system. Revisiting the same methods of sexual gratification, or taking the same series of turns through the same sequence of traffic lights. Play is a method for boring through boredom in order to find and experience the deep nature of ordinary things in the universe.

My daughter's "step on a crack" game clarifies things, because she serves the role of both designer and player: when racing through a crowded and noisy shopping mall, how can the vertiginous experience of being a small person pulled through the wakes of larger people be transformed from nuisance into advantage? By disallowing arbitrary footfalls, by constraining steps within the boundaries of tiles.

When enacted, the result produces an experience not only *different* from being whisked willy-nilly through the crowd, but also *larger* than that experience. Suddenly, a series of actions that might otherwise have gone unnoticed become central to her attention: it matters where every step lands, even if the game has no human or machine referee to police the system.

The process of play is one of identifying, manipulating, and responding to those components. Rather than lament the dissatisfaction of being at the mall, of being a small person among giants, and of being subjected to my schedule and my errands, my daughter identified, embraced, and manipulated the materials that surrounded her at that very moment:

> The speed of her own forward momentum, as my large hand and adult gait pulled her faster than she was able to keep up, but which connected her to a pace and a direction that she did not and could not choose.
>
> The patterned, rectilinear tiles of the mall floor, large enough that her small feet could fall fully between their edges and large enough that movement between them would require a leap more than a step.
>
> The crowd of other people steering around us as we steered around them, most of them legs and feet from her vantage point, and therefore easy to cast as generic obstacles rather than individuals.

Play entails a paradox: it is an activity of freedom and pleasure and openness and possibility, but it arises from limiting freedoms rather than expanding them. The boundaries of a playground, the contents contained within them. Their structures. Colloquial senses of game, play, and fun would hold that these activities amount to going outside the boundary of normal behavior, of *doing whatever you want*: "Don't play with your food" or "stop fooling around." But in fact the opposite is true: interesting play experiences arise from *more* constraints rather than fewer. Erecting barriers and boundaries more clearly delineates a system. As Bernard Suits puts it, play requires its participants to accomplish something "using only means permitted by rules, where the rules prohibit more efficient in favor of less efficient means, and where such rules are accepted just because they make possible such activity."[21]

Normally, we address a play experience like my daughter's either as if it were separate from the trip to the mall or as if it were perpetrated in the service of the errand I had dragged her along on. On the one hand, we could construe my daughter's activity as a distraction from the "real" work of running errands and therefore existing outside the domain of mall going, a play activity meant to release the boredom and unrest of being somewhere unpleasant. On the other hand, we could see her improvised play as a welcome and even a necessary distraction to help facilitate the rest of the afternoon's errands. The first interpretation assumes the work-play differential—that the work of chores exists nearby but orthogonal to the play that would divert a young child from boredom. The second interpretation invokes the productive repurposing of play as a means to pleasure or sanity, a resource to be put to use in the interest of "real" effort.

The truth is stranger than either option. My daughter's game isn't a distraction from errands, nor is it a mechanism to make errands possible. Instead, it's an activity *made out of errands* and other things too, like legs and ceramic tiles, in the same way golf is made out of grass and sand and rubber and wood—and leisure and wealth and zoning. A playground.

While Huizinga's examples of the play element in culture are far weightier than shopping, the profundity of war and politics and the like can hide the ubiquity of play. Play isn't only an activity whose surprising uses can be found in serious, consequential activity. It is also a condition of the world, everywhere we look, available to us should we choose to see it—and even if we don't.

PLAY IS THE WORK OF WORKING SOMETHING

An apparatus or experience fashioned by the boundaries of a magic circle is not necessarily a "game" or a "toy." After all, a highway system or a family budget has as many constraints as Monopoly or Super Mario Bros. Instead of calling everything a game, we should think of everything as *playable*: capable of being manipulated in an interesting and appealing way within the confines of its constraints. *All* media are playable when we look at them in the right light. And that light need not entail the total reform of our educational system, as Gray implies, and it need not signal the resolution of insufferable institutional autocracy, as Sicart suggests. Rather, play is the work of operating a subset of the world, one separated from itself via the circumscription of the magic circle.

The playground offers another perspective on the ironoiac madness of its mirror image, the protective encasing symbolized by the plastic sofa cover. By enclosing and encapsulating objects of experience, irony protects us from them, but in so doing it removes them from possible experience. Malls and school and food and everything else are transformed into motifs rather than cohorts. The best we can do with them is to emblazon them on T-shirts or tweets or Tumblrs, to use them as catalogs of insufficiency, bestiaries of lost opportunities.

On first blush, ironic circumscription looks similar to tracing the magic circle, to erecting a playground. A thing arises, cheeseburger-flavored Pringles or espresso machines or a Frisbee, and it preoccupies the attention of its observer. That thing is isolated, then contained within the security of irony's seemingly impregnable blister pack. Inevitably, irony's makeshift prison doesn't hold, and both the object and its prison prove untrustworthy, demanding new enclosure. And irony's gambit continues on ever larger scales, never offering succor but only increasingly larger and more cumbersome enclosures.

But there's a difference between the ironic and the playful circumscription. The former holds the object at arm's length—beyond arm's length, really, far enough to defuse it as a threat, and in so doing shields the ironist from all possible encounter. Irony is the playground circumscribed but then abandoned. Encounter leads to potential disappointment or betrayal. Better to treat everything as a threat, to trust nothing, to experience nothing save involvement with distrust itself. Confinement, regulation, and control characterize ironic circumscription. And like all good prisons, the ironoiac's object of mistrust is caged, isolated from its warden so to hold its savagery at bay. Who knows what potato chips and roadways and lawns might do if unleashed?

By contrast, the playground includes the observer as a member. It fashions the would-be ironist into a participant—whether as operator or observer—but still maintains the tenuousness of that involvement. Like a chalk line on pavement, the playground knows that it is arbitrary and temporary, flexible and negotiable. Play takes ironic detachment and transforms it into the conditions that bring about the experience it makes possible. Play refuses to presume that the golf course or the shopping center is reasonable or even desirable, a legitimate and certain source of basic operation, let alone success or meaning or joy. Instead, it merely asks what might be possible when things like fairways and malls are encountered by human agents.

The playful stance is the opposite of the ironic one: an embrace of the thing in question rather than a rejection of it. But not because play is more earnest or sincere, and not because it represents the free and liberated will of the player, whose volition elects the play experience instead of some other, less desirable labor or chore. For the ironoiac, the threat of an object's insufficiency produces paralysis. But for the player, this insufficiency is assumed from the start, as a necessary condition of play. Play is impossible without restriction—not doing what you want, but determining what is possible to do given the meager resources provided.

Play and the resulting effect we call fun are not loose human actions and emotions. They are created in conjunction with external objects. To experience fun, we must shift our reference from the joy or enjoyment we have come to expect from play, and instead understand play as a condition of objects and situations. Things are *at play* more so than we play with them. And when we encounter things that are subject to play, we need not subsume them into the domain of toys or games or playthings or other mere amusements, but we might instead simply allow them to be exactly what they are. Indeed, perhaps viewing them as what they are is the *only* way we can truly allow objects, situations, events, people, communities, and anything else to produce pleasure. Not by subsuming or capturing them, and not by deluding ourselves into believing that we are exerting our own control, creativity, and disruption over them. But rather by addressing each thing for what it is, while all the while acknowledging that anything is not entirely ours to address in the first place.

If fun is an admiration for the absurd arbitrariness of things, play is the process by which we arrive at that respect. Play is an activity, but even more so it's a material property of all objects—from guitars to steering columns to malls to lawns to language to, well, games—and fun is a sensual quality that emanates from them when we touch these things in the right way. Discovering, choosing, managing, and living with what's inside a particular playground—that's where fun, and where meaning, resides.

FIVE

From Restraint to Constraint

*Fear and irony make asceticism seem like the path to happiness.
But the world is too replete to reject like monks or saints.
Instead, we must embrace the constraints things impose upon us.*

REMEMBER THE KETCHUP TUBS AND DEODORANT SLEEVES we met at Walmart? If you've ever shopped for products like these, you know the terror of selecting one among the dozens of flavors, scents, shapes, and sizes available. The psychologist Barry Schwartz has a name for this familiar problem: "the paradox of choice."[1] In theory, the more choices we have—the different sizes and brands of ketchup, say—the more content we ought to be, because we can exercise our autonomy to select the option best suited to our lives.

But Schwartz argues that just the opposite happens in practice. Faced with so many choices, we only become more anxious about selecting the right one. When you realize in retrospect that a particular jug of ketchup was too big or too small for your needs, or that a particular scent of deodorant didn't quite match your sense of musky self-identity, you have only yourself to blame. After all, you could have made better choices.

To escape the paradox of choice, Schwartz recommends reducing available choices and focusing on what decision theorists sometimes call *satisficing*—finding and accepting the best option available. He offers methods to do so, organized around setting goals, assessing options, and making choices that help you tolerate the best scenario under the circumstances.

Back in 2004, when Schwartz suggested literally reducing our available choices as a mechanism for reducing anxiety, it seemed somewhat preposterous. In a world of Walmarts, how could one even lead a life of fewer options? But today, such reduction techniques abound. An Irish man lives without money through scavenging and freecycle.[2] A Texas college dean lives in a converted dumpster.[3] The Japanese organizing consultant Marie Kondo's runaway bestseller *The Life-Changing Magic of Tidying Up*, a book about how to purge your life of superfluous belongings, reportedly has led to a massive uptick in charitable donations as tidy-uppers clear their closets. Toss your couture and don your habit—asceticism is hot.

But is abstaining from material goods really the answer to the paradox of choice, or is it a convenient affectation for those with enough material advantage to choose? Formal, public declarations of self-imposed austerity are on the rise. In one such specimen, the *Wall Street Journal* writer Katy McLaughlin congratulates herself on a new "two glass a week" wine diet, a salve for a life of "self-indulgence" she and her husband had settled into. But reading McLaughlin's description of the problem makes it seem like the paradox of choice isn't really the problem that plagues her. Instead, a new obsession with satisficing at all costs:

Maybe it was when we left New York City four years ago and daily life just got easier. Maybe it was when Alejandro's career picked up and we had more disposable income, or when the kids started school and didn't need pricey baby sitters anymore. Whenever or however it happened, over the past four years, we began indulging ourselves in ways we could never have imagined when we were young and broke.[4]

"Our self-indulgence gave me pause," she reports having worried, before describing the regimen of reduction she imposed on herself in order to make wine tolerable again.

Imagine reading these words through the eyes of a family in the Great Depression, or a college student starting her first job, or even an upper-middle-class parent with an infant at home. Think of how outrageous it would sound. Somehow, for Katy McLaughlin, the experience of enjoying wine she can easily afford to drink in healthful moderation has become an "indulgence" that must be tempered. Without an artificially imposed reduction of supply, wine becomes intolerable. No consumption except as a counterpoint to abstinence. Let me not eat cake!

Our fear of things isn't limited to cloying, ghastly, mundane products and experiences. Even the very dreams that time and material comfort make possible become subject to ironoia's affliction. Regionally sourced, estate-bottled wine is as threatening as Heinz 114-ounce pour, store, and pump-jugged ketchup, despite the former's higher social status. It makes you wonder: what good is satisficing when the problem isn't only in choosing, but in anything whatsoever one might choose?

• • •

SOMETHING ESSENTIAL HAS changed since Schwartz suggested goal-oriented satisficing as a solution for dissatisfaction: we now spend more of our time with computers than we do without them. According to Nielsen data, Americans use up 70 percent of their free time with electronic media—television, smartphones, the Internet, video games, and so on.[5] And in the digital world, the material constraints of experience don't apply. One need not be able to afford a bottle of wine every day to read Katy McLaughlin's article about home oenologics or the many ripostes to it one could find online. One not need even drink wine to peruse Pinterest boards full of decanters or recipes or "kitchen hacks." Today, you can live a million lives all in one afternoon on your smartphone. No wonder we're all so exhausted.

We burn out one another with them, too. If you're one of the two billion or so people on Earth who own a smartphone, you're probably a jerk with it at the dinner table or the conference table or, really, anywhere else for that matter. Maybe it's buzzing with notifications and messages, begging you to pull it from your pocket or pocketbook. Even if you try to ignore it, even if you want to, it purrs and whimpers like a kitten until you finally give in and reach for it, thinking you're doing it a service by alleviating its suffering. Just a glance, you tell yourself, just to see if it's important.

But soon enough, hypothetical importance gives way to mere nearness. Like cancer comes for free with cigarettes, ironoia comes for free with smartphones: in any situation, they offer a reminder that something better might come along, and so they contrive a reason to mistrust whatever appears as figure in favor of whatever else might be hidden in ground.

Sometimes we mislead ourselves into thinking that the cultural tide has shifted, the draw of the smartphone serving as a Narnian portal to magical elsewheres for the Millennials (and younger!) who cling to them, while we olds shake our heads in disapproval or despair, no less subject to the smartphone siren's buzz, but at least aware that we ought to be tying ourselves to our proverbial masts. I don't want to admit how many times I've picked up my smartphone, tousled it, and put it back down over the course of typing the last few paragraphs.

Critics like the MIT sociologist Sherry Turkle have argued that our electronic devices are taking us away from face-to-face interactions—she calls conversation the most human and humanizing thing we do—but this old-fashioned, polite society dream is the equivalent of Wampole's call for earnestness as a salve for irony.[6] The truth is much weirder. Smartphones even distract us from our smartphones; one notification from a background app suggests that it might have something better to offer than the currently open app.

Everyone oscillates between loving and loathing their devices, young or old or even younger. In 2012, a twenty-year-old preschool teacher named Stephie devised a game to try to tame the glass lions in our pockets. You're meant to play when you dine out with friends. After everyone has placed their order, all the phones go onto the table, stacked facedown on top of one another. Until the meal ends and the check is paid, the phones remain stacked, no matter how much they purr and whine, demanding your attention. If someone is overcome and reclaims a phone and its concomitant data-giving pleasures, so be it—but that person must pay for the entire group's meal. Stephie originally called it "don't be a dick during meals with friends," but the many news organizations that reported and commented on it redubbed it the "phone stack game."[7]

Turkle would applaud Stephie's tactics. Her game is one of an endless supply of willpower tricks that have become popular as a way of staving off our surplus of constant options and opportunities. But the problem isn't smartphones and other digital tools. The problem is having so many options, all of which could easily be cast aside for so many others, human, digital, or tangible. Ironoia applies to the people in the room as much as it does to anything else.

...

TURKLE'S PLEA FOR presence is the latest in an endless procession of moral codes. Once, virtues were handed down to us, often from a deity or a religion. Allowing a higher power and a millennia-old institution to dictate behavior on your behalf is an effective way to outsource the problem of choosing right from wrong. But in our secular age, we even have innumerable options for moralism! Every talk-show guest and nonfiction book and Internet think piece offers a new take on what you should embrace and what you should abandon.

Life today is full of directives and rejoinders, all reliant on the victory of the self over itself. The things we're meant to do usually amount to mandates that we, as individuals with some control over how we behave, aspire to follow. Psychologists have reframed willpower as a biological and neurological phenomenon rather than a virtue, and as with all matters neurological today, that means our willpower can be managed or controlled, whether through mindfulness, therapy, prescription drugs, or (eventually) cognition-enhancing technology.

Directives and rejoinders and guilt and anxiety existed long before the secular, cognitive worldview became predominant. But the milkshake of secularism, neurophilia, individualism, and consumer capitalism have made *you* the center of the things you do—or ought to do. Eat less. Exercise more. Drive less. Read more. Recycle. Buy local. Go out. Stay home. Spend less. Save more. Your waistline or health or bank balance become a function of your knowledge about and willingness to *restrain* yourself. Skip dessert, and sock that paycheck away instead of splurging on the steak dinner. Put the smartphone away and talk for once. Be more *mindful*, for Buddha's sake!

Living this way involves a constant struggle between what we really want—or what we think we really want—and what we ought to do instead. As a result, we feel either guilty (why did I eat that slice of cake?) or disappointed (I wish I'd eaten that slice of cake!).

Those feelings are Barry Schwartz's paradox of choice rearing its head—once the opportunity to choose presents itself, the cards stack themselves against you. The likelihood of the universe configuring itself to correspond with your specific desires is already low. But even worse, your desires are likely to reconfigure themselves in the interim. Or the conditions for those desires are likely to change. Sometimes these shifts happen slowly: eggs, once considered a source of high cholesterol that should be avoided, suddenly become a tool to reduce overall cholesterol, thanks to being low in saturated fat. Sometimes they happen quickly: someone already took the last slice of cake, anyway. Or posted a photo of it on the Internet.

And yet we continue to look inward, asking ourselves what we really want. What do you want to write or eat, where you want to visit or work? Most often, those questions lead you to other questions: what story do you want to tell, or what food is best to feed yourself or your family, or what kind of job would you enjoy or be good at or at least not hate as much as your current one? Even the most mundane kinds of choices can quickly spiral out of control. Should you paint the bedroom Tibetan sky or faraway blue? Should you plant herbs or vegetables? Should you watch a documentary or indulge in an action flick? Should you drink wine twice or only once per week? Any question seems two degrees away from total existential uncertainty: Who am I? Why am I here? What does it even *mean* to want something, anyway?

We remain uninspired by the chance to pick a paint color or plan a trip or find a job, even if we simultaneously want the results to delight us, or at least satisfy us, or *at least* not to harm us. And so we're constantly tempted to fall back into the dubious comfort of irony because it allows us to avoid facing the problems that decisions pose. Irony seems like the best option because it is the safest one. Rejecting experience is the only way to ensure that it cannot bite you in the brain for having encountered it.

Irony descends from restraint and amplifies it. "I shouldn't . . . " becomes "I won't, but I can have it both ways." Instead of partaking in *or* rejecting cake, one can fashion Pinterest boards of hypothetical cakes one is sure never to eat; or indulge in the delightful (if mortifying) disasters of the website Cake Wrecks, where malformed or poorly decorated cakes allow us to laugh at the supermarket bakery underlings who fashion them; or watch others spend thousands of dollars on cakes that don't get eaten on television shows like *Ace of Cakes* or *Cake Boss*.

THE NEW ASCETICISM

Imagine that you want to do something. It doesn't matter what: paint a bedroom, read a novel, make a Sunday dinner, take a vacation, get a new job, or look for a good television show. How do you begin? Schwartz suggests that we should start by having fewer options in the first place. The surplus of paint colors at Sherwin-Williams, the infinity of programming on Netflix, and the wealth of possible travel destinations near and far only ensure anxiety and disappointment. By reducing opportunities, Schwartz argues, not only do we fashion an excuse for taking the best option available without worry, but also we allow the resources that would otherwise produce more choices to benefit the greater public interest.

As a description of a problem, the paradox of choice is a powerful diagnosis of the bittersweet success of modernity. Schwartz's book is subtitled "Why More Is Less," and he offers a compelling account of why the massive number and variety of options at every turn paralyzes more than encourages us. But he offers little advice about what to do next, apart from suggesting that individuals learn to accept satisfactory solutions, and insisting that sharing prosperity across human societies will also reduce the cognitive burden of choice. The former advice amounts to yet another call for restraint:

just desire less! And the latter advice amounts to an unachievable fantasy of global socialism.

The times have changed, too. Back when Schwartz invented the paradox of choice, David Foster Wallace's television-era irony had not yet exploded into our present-day Internet-stoked surplus. Capitalism notwithstanding, the idea of calming the anxiety of choice by reducing the supply of choices might work in theory for shampoos and mustards and automobiles and schools, but it offers no succor against the Cambrian explosion of symbolic objects that litter contemporary life.

Schwartz assumes that material scarcity imposed through policy or willpower can spare us from the burdens of choice by divesting us from the sheer volume of goods and services about which we might make choices. I suppose we can imagine, unlikely as it might be, a hypothetical, collectivist future in which the wealth created by the surplus value of limitless variation in condiments and underpants and bank loan products is reinvested in greater global access to these and other products. But such a world wouldn't really help even if we could achieve it. There's no reason to believe that the burden of choice would be freed from the grip of ironoia, where an infinity of hypothetical, imaginary mustards and mortgages can be divined out of nothing and reanimated into material form on Snapchat and Instagram. We'd still be bored for not having pursued our leisure opportunities fully, and we'd inevitably seek solace by ironizing the new comforts we haven't yet invented. Restraint is always temporary. Like cleaning up by hiding things in drawers and closets, it merely defers mistrust into the future.

As we already discovered in the context of play, the autonomy we nickname *freedom* amounts to imprisonment. It challenges us to aspire for something other than our reality: it wants you to work harder, so you can earn a promotion, so you can earn more, so you can afford a better house, so you can shorten your commute, so you can spend more time with your family. But sure enough, even if such aspirations are realized, they lead only to greater aspiration: that new job demands even harder work with even longer hours, erasing the time won back from the commute, not to mention making family time even more elusive. Meanwhile, your new, better house only made you more aware of the newer, better houses even closer to work, the ones you still can't afford.

As a strategy for living better, restraint feels natural, partly because its values are deeply ingrained in Western culture. Restraint connects to the Judeo-Christian traditions of guilt, good works, and reward that also found the basis of the global economy. As a result, restraint appears to exercise free will and autonomy, facilitating a benefit we value more than any other: freedom. But if you stop to reflect on the feeling of restraining yourself, it's a terrible sensation: that of constantly *denying* rather than accepting your circumstances, of constantly second-guessing your choices and then lamenting all the others you could have made instead.

■■■

SO URGENT HAS our mistrust of things become, we have begun to embrace restraint at any cost. Asceticism is in, and it's glamorous. Marie Kondo, a Japanese organizing consultant, has sold millions of copies of her books on decluttering. Her "KonMari" method: by putting your belongings in order, by ridding yourself of unnecessary items, and by conducting a holistic bout of decluttering rather than a space-by-space or room-by-room one, you can achieve a lasting contentment.[8] KonMari recommends the depopulation of your living space, and Kondo offers a singular criterion to determine if items ought to stay or go. The successful tidier, according to Kondo, is one who is "surrounded only by things they love."

The process of removing things to create completeness has a history in Eastern culture. The principle of *yohaku* roughly corresponds with what we'd call "negative space" in the West. It finds its way into Chinese and Japanese art in the sparing application of ink or pigment on drawings and paintings, or the spartan design of Zen gardens. Western modernism holds similar values; for example, the maxim of "less is more" that Schwarz inverts in his subtitle comes from the mid-century architect and furniture designer Ludwig Mies van der Rohe. The French writer and aviator Antoine de Saint-Exupéry fashioned a less austere and more poetic version of the same idea: "perfection is finally attained not when there is no longer anything to add, but when there is no longer anything to take away."[9]

Kondo's workaday version of yohaku is more pragmatic and less pretentious: you start KonMari by discarding broken things, obsolete tools, and once useful objects whose utility has since passed due to changing circumstances. As for the rest, Kondo offers one sole heuristic for determining whether an item should stay or go. "Take each item in one's hand," she urges, "and ask, Does this spark joy? If it does, keep it. If not dispose of it."[10] Kondo admits that it's a vague guideline, but she trusts your ability to make an affective diagnosis. "The trick is to handle each item. When you touch a piece of clothing, your body reacts."[11]

Like Schwartz and Wampole, Kondo's approach to regulated asceticism puts the burden of creating conditions of simplicity on the agent who would presumably benefit from its outcome. But for Kondo, this purge is a one-time process. After the hard work of an initial purge, the temptations of modest hoarding to which we are all susceptible can be overcome, because Kondo encourages tidiers to anthropomorphize their skirts and appliances and handbags and razors and all the rest. Her approach attributes a soul or a life to inanimate objects.

This commitment to animism (a tendency more common in the East) cuts both ways. On the one hand, it gives KonMari an easy model for practicing a respect for the objects with which its practitioners surround themselves: think of them like kittens or hedgehogs or other living creatures. Recounting a particularly egregious act of contempt against socks by a client, who had rolled them into balls for storage, Kondo exhorts, "Look at them carefully. This should be a time for them to rest. Do you really think they can get any rest like that?"[12]

But on the other hand, anthropomorphism puts us at the center of the universe. The socks are judged worthy of retention not by virtue of their provenance, their craftsmanship, or another intrinsic feature, but based on how they make their owner feel. Objects are possessed, and their human owners determine the fates of those objects by passing subjective judgment on their ability to yield comfort. Once more the world is held at arm's length, and the self embraced. Mindfulness over worldfulness. And for exactly this reason, Kondo's intervention has been enormously successful. After the US release of *The Life-Changing Magic of Tidying Up*, secondhand stores like Goodwill reported substantial boosts to their inventories as stuff-weary Americans unloaded their joyless belongings.[13]

THE NARCISSISM OF RESTRAINT

Joy is relative, I suppose. George Carlin famously suggested that to you, your stuff is stuff, while others' stuff is shit—"Get that shit off of here so I can put my stuff down."[14] But KonMari's gentle exterior betrays its inherent narcissism. Egocentrism is a danger of most appeals to restraint. Like irony, restraint has its origins in the self rather than the other. Giving away your stuff might successfully short-circuit the dread of sincerity-or-contempt, but it also outsources the work of determining an item's worth to unseen machinations and underlings. We get to pat ourselves on the back for taking account of all the waste we produce and consume, but then we relegate its disposal and future management elsewhere, as someone else's problem. "When we flush the toilet," the ecological philosopher Timothy Morton analogizes, "we imagine that the U-bend takes the waste away."[15] But there is no "away"—there's just more world, and your waste spills over elsewhere, becoming someone else's problem.

Still, it's no surprise that the urge to restrain tugs at us. Even in Kondo's closets made into Japanese gardens, the ambiguity of joy introduces new doubts. Whose joy, and when, and under what conditions? As the writer Carlye Wisel asks of KonMari, what if *all* your stuff sparks joy? "To me," she says, "joy is opening a closet and seeing things you currently love, once loved, or do not yet know why you purchased, and mix-and-matching them into an outfit that conveys your current mood at that moment in time."[16] Socks and dresses and plates and tchotchkes do not enter into singular relations with you or me or one another, but extend in space and time, encountering other companions even after they exit the proverbial U-bend of Marie Kondo's reject piles.

The desire to issue restraint marks another symptom of ironoia. When something offers promise but also threat, as do closets and cake, it's far easier to reject them and then pat ourselves on the back for doing so than to engage with them more deeply. Perhaps all exercises of restraint are actually ironizing acts, symptoms of our fundamental boredom: holding something at arm's length, refusing it, and disposing of it all fashion a phantasmal copy, a ghost that we can then pride ourselves on having spurned. "Well, *I* didn't eat any cake."

Contemporary life is both intensely consumerist and increasingly virtualized, so it's no wonder that restraint's stock has risen. Think about your laptop or your smartphone. Is it a source of joy? If Marie Kondo forced you to subject it to KonMari's acid test, would it be found wanting? The truth is, it's complicated. Digital devices do so much now that it's hard to know how we could separate their promise from their threat. A smartphone is a source of connection, of companionship, of information, of leisure, but also of distraction, of compulsion, of disconnection, of obsession—a whole microcosm is contained amidst its glass and aluminum and silicon and software. Like Wisel's closet, the smartphone contains multitudes.

At times, of course, the smartphone represents the worst kind of mistrust of things, the sensation that we cannot control ourselves. One definition of addiction is wanting to stop a behavior but not being able to do so. In that sense, the stream of updates and notifications from Facebook or Twitter or Instagram offers delight and discomfort all at once. Now that so many jobs are knowledge work, and now that so much of our leisure takes place digitally—social media, games, streaming music, reading, news, movies, and television—working can give way to puttering with zero switching cost.

Consequently, a whole new category of tools meant to help us manage digital distraction has arrived. Special-purpose word processors like WriteRoom and Ommwriter offer Japanese garden–like serenity in full-screen, supposedly Zen-inducing environments that block out the wayward web browser and e-mail client. A utility called SelfControl can be set up to block access to websites and mail servers for a specified time. Another, called Freedom, disables a computer, tablet, or phone's networking capabilities entirely for a specified period (if you need to cheat, Freedom's creators reveal, you'll have to go to the trouble of rebooting).

Unsurprising, to me at least, most of the testimonials that appear on Freedom's website are from writers. "If I ever finish my book, this is why," quips *No Logo* and *Shock Doctrine* author Naomi Klein.[17] Some take matters even further. The technology critic Evgeny Morozov locks his electronics in a timed safe when he writes—on another computer, whose Ethernet port he plugged and whose Wi-Fi card he forcibly extracted.[18]

Even though these cases seem like exercises in suppression and control, the flipside of omission is inclusion. It's a distinction more easily understood from the perspective of play. By excising the Internet-connected features of the computer, its user is able to focus on a different aspect of that general-purpose machine—writing or programming or graphic design or computer-aided drafting or whatever. That the software is called "Freedom" is no accident—it creates one set of liberties by removing another. But unlike Kondo's method, which tidies by permanent extraction, Freedom, WriteRoom, and Morozov's safe admit that there is pleasure in both connected and disconnected computational experiences—but that each of these two pleasures might not be possible without erecting a firewall between them. They represent different circumscriptions of computation, different digital playgrounds, purpose-built and inhabited for different reasons at different times.

Using restraint to reject some things in favor of others is not wrongheaded, but it does embrace two fatal errors. First, it promotes *rejection* rather than *attention* as the principle gesture. And then, it takes the result of that rejection as a definitive, permanent one. Katy McLaughlin wasn't really indulging and then suppressing her love of wine. She was erecting different magic circles, one for the routine of daily enjoyment and one for the anticipation of more infrequent fulfillment.

The erection of a playground is pliable and tentative. It rejects wine or cake not for the sake of rejection, but in order to preserve a sense of delighted contentment after a meal, or to observe a chosen diet, or to manage a chronic condition, or to keep a budget, or to save room for a secret outing with a spouse, or to create focused attention around a luxury that has become too ordinary.

Restraint ironizes a playground. It holds it at a distance, treating it as a principle rather than an experience, a pretense rather than a preoccupation. Often, the more attention a limitation is afforded, the more you can be sure that its purpose is an ironic affectation rather than a deliberate pursuit.

...

TAKE REAL ESTATE, which has been subject to a series of trends in extreme downsizing in recent years. Some of this shift responded to economic conditions. American housing starts since the 2008 recession show that median home sizes are shrinking, as home loans contract and as families return to cities from suburbs. But the modest reversal of the exurban McMansion trend of the last quarter of the twentieth century has been overshadowed by a new trend known as *microhousing*.

In contemporary Manhattan, tiny flats are extending the tradition of the studio apartment first invented in the 1920s to make big-city life affordable. Among them is Specht Architects' glistening modernist retreat, which occupies only 425 square feet on Manhattan's Upper West Side. The flat boasts an ingenious use of vertical space, with intricate hidden storage. Big windows flood light into the small apartment, whose dark wood floors and vaulted, white walls and ceilings make it feel bigger.

But, is it really affordable? *CityLab*'s Kriston Capps explains that the resulting microloft costs $950 per square foot.[19] For comparison, the US national average cost per square foot for residential home construction is $125. The most expensive median price per square foot, found in California, is still only $256. Microhousing, it would seem, is delightful for residents of densely populated cities who can afford it—and who don't mind giving up private space for the spoils of one of the world's great cities right downstairs.

But couldn't lessons from Manhattan find their way to smallville? Even in cities and suburbs where land is less scarce, more and more people are rejecting excess and choosing to build tiny microhomes of only a few hundred square feet. Or so goes the story, anyway. Reality turns out to be more complicated.

Specht Harpman also offers a prefabricated, 650-square-foot option at about $450 per square foot—still almost four times the national average. The design is striking: two cantilevered rectangular prisms, complete with solar panels, water reclamation, and organic waste diverter. But as Capps notes, even after shelling out $300,000 for a prefab home, you still need a plot of land to put it on. Land is cheap off the grid (American farm and pastureland costs between $1,500 and $3,000 per acre, but seaside or mountain land costs more), but when all is said and done you might be better off building a traditional home, at least if cost is the major consideration. And even ignoring cost, a 400- or 600-square-foot home, whether on the Upper West Side or in Upper Michigan is mostly useful for a single professional or a childless couple, not a family. Microhousing doesn't really involve a return to the modest housing practices of the mid-century. The average, new single-family home in 1950 was 983 square feet, at an average cost of $11,000, inclusive of the land on which it sat. That's about $109,000 in 2016 dollars.

The word *microhousing* insinuates modesty and restraint, but the reality doesn't embrace those virtues. Here, restraint acts as a fashion more than a life philosophy—a "fetish object," as architectural writer and *Dwell* magazine founding editor-in-chief Karrie Jacobs calls microhomes.[20] Of course, there's nothing wrong with fashions and fetishes. The problem comes from deluding ourselves about what materials the microhome playground really circumscribes. The $950-per-square-foot Manhattan loft or the $500,000 off-the-grid microhome struggles to comply with reductions in financial and ecological impact. Instead, microhomes become luxury goods.

Instead of seeing microhousing as an exercise in greater modesty, it's better to understand it as an intriguing exercise in constrained architecture and residential life. They ask what sorts of living spaces can be architected within a limited area, and then they ask the occupants of such spaces what kinds of lifestyles they might be able and willing to lead within them. By reframing the micro-home as the result of a playground erected by architects, planners, and homeowners, the true nature and purpose of these structures can be better understood and practiced. The same is true for all exercises in restraint. The moment you hear someone championing pure asceticism, they're probably just expressing anxiety.

CONSTRAINTS CREATE ABUNDANCE

It may seem strange to talk about dwelling with language usually used to describe the arena for a game. The sickly sweet Mary Poppins idea that everything can be a game might tempt your inner contempt to erupt. Surely architecture, building construction, homeownership, and the daily trials and pleasures of dwelling aren't *merely a game*?

If it bothers you, think about play instead. We experience games by "playing" them, and play is an activity we tend to associate with freedom, with being able to do whatever we want. This view of play stems from conditioning ourselves to see play as the opposite of work (or school or chores or obligation more generally). If work is what we *have* to do, play is what we *get* to do. But this attitude is mistaken, a consequence of seeing freedom as an *escape* from imposed restrictions rather than as a practice of working within adopted constraints.

In fact, games and play offer the opposite: an invitation to do *only* what the system allows, for no reason other than the fact that it was designed that way. Games are built out of constraints, and play arises from limitations. Paradoxically, if we add limitations to an experience deliberately and knowingly, the more interesting and appealing that experience tends to become.

Here's a simple example, a game you might have played on road trips as a kid called "I'm going on a picnic." There are many variants of this game, but in the one I remember the first player says "I'm going on a picnic, and I'm going to bring . . . " and then names a food item that starts with the letter *a*, say, "apples." The next player does the same, repeating the *a* item and then adding another item that starts with *b* ("I'm going on a picnic, and I'm going to bring apples and bananas"). The game continues with as many players as desired, each repeating the sequence and adding an item. If a player cannot remember part of the sequence, or cannot name a food appropriate for a picnic, he or she is "out."

This game is simple, even trite. But in its simplicity, it explains the act of circumscription central to play. First, its players fashion an arbitrary conceptual space, creating a playground made of words and turns. But then, they treat the resulting materials as far more reasonable than they deserve to be treated on their own, using the excuse of the magic circle as a rationale for ignoring its arbitrariness, for exerting the defamiliarizing action that turns a previous ground into a new figure.

It teaches a surprising lesson: the *more* constraints we add to the experience, the more interesting the result feels. Imagine a game in which each player simply said a word in order. That's the prototype for "I'm going on a picnic," after all. Such a game might distract a toddler for whom even knowing words is still a challenge, but it would hardly capture anyone else's attention (unless they were to play it with a toddler, of course). "I'm going on a picnic" takes that structure and adds limitations: each word must be a food that could be taken on a picnic; each word must begin with the next letter of the alphabet in sequence; each player must remember the entire previous sequence before adding a word.

The result is reasonably compelling, but still not very challenging for older kids or adults. Increasing the appeal and depth of the game for such audiences is certainly possible—just add even more constraints. For example, instead of choosing foods in alphabetical order, each new food item could be required to begin with the last letter in the previous word ("apple" begets "egg," then "grapes," and so on). Such an addition makes recalling and reciting the sequence more difficult, and it also prevents players from planning moves in advance. Suddenly, a silly, forgettable activity—reciting words in order—becomes a compelling experience that warrants serious attention. The experience becomes much *larger* than the constraints that create it.

By embracing more limitations, a seemingly meaningless idea becomes a more meaningful experience. This paradox of play—the idea that fun arises from limiting freedoms rather than enhancing them—isn't only true of board games or card games or playground games or video games. It can be found in any kind of material whatsoever.

If the imposition of external *restraint* hasn't been effective, why not embrace its opposite: *constraint,* the adoption of controls and limitations from *inside* rather than *outside* a situation. Constraint has the flexibility to cover all forms of material construction across all media. Rules describe the internal logic of a system, but constraints delineate its edges, the membrane that contains these machines and separates them from other beings, creatures, devices, and experiences in the world. Constraints are the features that delimit both the system's characteristics and the user's possible actions.

Constraints inspire us to see something like microhousing as a kind of architectural play rather than a type of humble living. Thinking of tiny homes as domestic playgrounds demonstrates that restraint is not a useful tool for creating meaningful experiences. After all, you don't need a micropad to explore the lifestyle possibilities made possible by the architecture of your home or apartment. Because we are obsessed with restraint, we think that reduction always leads to greater gratification, as we shed our shameful hunger for more of everything. But in truth, more or less of something doesn't make it meaningful, valorous, injurious, or boring. Rather, the contents of a specific circumscription gives value, and how deliberately and seriously one treats the results.

■ ■ ■

WHEN I FIRST moved to Atlanta from Los Angeles, I was eager to take advantage of the far more modest real-estate market and buy a house. Restraint or not, even with a fairly lucrative tech-industry salary, Los Angeles's real-estate market was well out of reach for our family, unless we wanted to commit to a long, traffic-addled commute. Having lived in another car-oriented city, I knew we wanted to live in town and near mass transit. Part of the trade-off for homeownership was making do with one automobile—not a burden, but a bit of a trick in a city like Atlanta, where the expansion of rail transit had been stymied by complex multicounty politics and decades of white flight to the exurbs.

Such a situation represents *opportunity* as much as restriction, but understanding it in such light requires circumscribing *oneself* inside a playground that also contains these other limits. It's not a novel act; people do it all the time. Anyone who has ever searched for a house or apartment, shopped for groceries within a budget, planned an afternoon and evening of errands, or conducted just about any business whatsoever in the world already understands the secrets of constraint. But what we often miss is its flipside, the freedom of movement created by defining a space in which to think, dream, plan, observe, or act.

We ended up with a cute, brick bungalow in a nice part of town, within walking distance of the train. It had been built around 1950, and offered roughly the 983 square feet that would have been common that year, spread across two bedrooms, living and dining rooms, a kitchen, and a full bath. In the 1990s, the previous owners had expanded the attic into two small bedrooms and added a dormer out the back to contain a second bath. The addition increased the living space to about 1,300 square feet, which is roughly equivalent to the average, new American home in 1970.

But as our family grew larger in individual and absolute terms, we began to run out of space. Actually, that's not true. After all, families of the 1950s had no problems living in exactly that amount of space—even less, since we were taking advantage of a former attic as two small kids' rooms. In the fifties, families would have had no qualms having two or three kids share a single bedroom. The cultural conventions and expectations for dwelling are what changed, both motivating and responding to the construction of more and larger homes—up to an average of 2,350 square feet by 2004, when we bought our little cottage.

To say we "ran out of space" is an ironic move, one that rejects the bungalow as insufficient or untrustworthy. In truth, something else happened: we made a conscious and deliberate effort to draw a new boundary for dwelling. Our circumstances were a part of this new playground. We could afford more, the real-estate market had recovered enough from the lows of the 2008 economic collapse (a phenomenon that was itself driven by immoderate speculation in housing growth), interest rates were at historic lows thanks to that economic calamity, our newly teenaged son had begun to knock heads with the pitched ceiling of his attic-conversion bedroom, and so on.

The house we now occupy is embarrassingly large, dwarfing even those 2004 standards that dwarfed our 1950s bungalow occupying 1970s living space. But the new playground of the dwelling affords new opportunities and new degrees of freedom made possible by exploring the play made possible by their material properties.

Enough kitchen counter space to permanently house the espresso machine that crafts the playground I described in chapter one. A guest room allows easier visits from family and friends, and a dedicated workout room has made the biggest difference in my ability to keep a commitment to working out regularly, because I am able to leave all the equipment and shoes and sweat in one place. I converted a lawn-equipment garage into a workshop, which I filled with a few power tools, facilitating a newly feasible devotion to occasional woodworking. The hobby's simple materiality balances the rest of my professional routine, which is unduly focused on pondering, using, programming, or writing about technology. My home is a delightful playground not because it's large, but because of the variety of things I am able to do with it.

I'll surely earn some sneers by seeming to gloat about the fruits of my material good fortune, hard-won or not. So let me be clear: larger houses with more rooms and more stuff are not the true path to gratification. Rather, the playgrounds we create for our experiences can vary in size and in quantity of contents, while still providing meaningful encounter. Restraint doesn't produce that meaning, since it represents the rejection of encounter rather than the adoption of specific, new ones. Constraint, by contrast, always requires a boundary, in which specific materials are contained and others are excluded. Constraints can thus entail very small sets of materials—living in 450 square feet or on $900 per month, say—or much larger sets of materials. Embracing constraint doesn't mean embracing asceticism. Rather, it means inventing or adopting a given situation as a playground in which further exploration is possible.

■ ■ ■

WE OBSESS SO much about our desires and dreams—on the ways things can produce satisfaction or alienation—that we seldom stop to look at the situation in reverse. Rather than asking whether our environment gratifies us, what if instead we asked whether we have yet explored the implications and capacities of the things with which we find ourselves surrounded?

This suggestion runs counter to what has become conventional wisdom in psychological studies of material goods and contentment. A principle known as the Easterlin Paradox (named after the economist Richard Easterlin, who proposed it in the mid-1970s) suggests that increased wealth produces greater happiness, but only up to a point.[21] Widely cited research by Daniel Kahneman pegged the point of diminishing returns at an annual household income of $75,000 on average (in 2010 dollars) in the United States—past that point, money doesn't make people that much happier (although that figure is subject to considerable regional variation).[22]

Building on the Easterlin Paradox, the Cornell University psychologist Thomas Gilovich argues that the material goods we buy make us happy for a time, but soon enough we adapt to them. The sheen of their previous novelty fades. To combat the diminishing returns of ownership, Gilovich and his collaborators suggest that we should spend money on experiences rather than things: travel, education, events, and so forth.[23] Speaking to the magazine *Fast Company*, Gilovich justified the value of experience over possession by making an appeal to incorporation: "You can really like your material stuff. You can even think that part of your identity is connected to those things, but nonetheless they remain separate from you. In contrast, your experiences really are part of you. We are the sum total of our experiences."[24] When the bar for valuing or respecting something is set by its ability to be made a part of our bodies, it's no wonder that we find ourselves so frequently faced with the terror of ironoia. Believing that comfortable coexistence with another entity boils down to *whether or not you can literally absorb it into your physical body* sounds more like sociopathy than psychological healthfulness.

The psychologists Darwin A. Guevarra and Ryan T. Howell offer a corrective to Gilovich and his kindred. Guevarra and Howell suggest that "experiential goods" like sports gear and musical instruments make people more, rather than less, gratified because they are acquired to facilitate activities rather than merely to exist as possessions or "pure goods."[25] The idea of experiential goods helps explain why both the Manhattan microhome and my decidedly non-microhome offer similar purchase on desirable experiences: both circumscribe a set of possibilities, even though both also demand a prerequisite outlay of capital. So too with kitchen appliances like my espresso machine (or gelato machine or panini press): the delight of owning such apparatuses comes not from their possession, most of which is spent stowed unseen in cabinets or atop counters, but in deploying them, in using them to make sandwiches and sorbets which in turn can be incorporated into the bodies of my family and friends, realizing the age-old custom of deriving affinity through breaking bread and stretching stracciatella.

But experiential goods still don't take things seriously enough. Guevarra and Howell name the obvious types of experiential good, the ones that correspond well with the wholesome version of Pier Paolo Pasolini's boyhood physical education: golf clubs and Super 8 cameras, pasta extruders and soccer balls. By this reasoning, items like strawberry and cream Bagel-fuls and Rainbow Brite toss pillows are relegated to the favela of "pure goods," objects acquired only through delusion, their existence forever fenced off from the languid meadows of experiential play.

But experience quickly gives way to rejection. Think about all the experiential goods in your own life that relegate themselves to disuse, despite your best intentions. The collapsible elliptical machine forever folded and stowed under the bed. The 4x4 SUV that you really prefer to leave garaged in heavy rains, let alone risk marring its alloy rims off-road. The posh set of golf clubs you bought to impress a client but never used after the trip was canceled. The dozens of shows and films forever saved to the purgatory of your Netflix queue. The thousands of articles sent to handfuls of read-later apps and services, only to languish there in oblivion.

Perhaps an object's potential for experience is not a function of its nature, but of our capacity and tolerance for treating it as a playable object, of exploring the constraints of the system it presents to us, of treating it *as an experience* rather than expecting it to create experiences on our behalf. As Boniakowski said of the Cheeto-as-haute-cuisine, so much of the joy of something lies in the solemn appreciation you bring to it. Who's to say you can't bring that solemn appreciation to cheese puffs as much as to cheese plates, to accounting software as much as to acoustic guitars? The question, of course, is *how* you might do so. The answer is easy to explain but harder to do: by taking things for what they are rather than for what you wish them to be, by acting on and with them despite the inherent absurdity and alienation in doing so, and to repeat this process over and over, continuously returning to the universe of things whose possibilities you thought you had exhausted, and giving them a chance to surprise you.

CREATIVITY IS CONSTRAINT IN CONTEXT

Artists and designers have long known that creativity does not arise from pure, unfettered freedom. Little is more paralyzing than the blank canvas or the blank page. At the same time, the creative process is not driven by restraint either; one does not paint a painting or write a novel or form a sculpture or code an app by resisting the temptation to do something else. Rather, one does so by embracing the particularity of a form and working within its boundaries, within its constraints.

The nineteenth-century English designer William Morris once said, "There is no art without resistance in the material."[26] He meant that art—or creativity of any kind—doesn't come wholly from within us. Things external to the thing created structure the creativity. By "material," Morris means the physical matter out of which art is crafted. A painting is made from pigment and medium applied to canvas. A poem is made from words formed into a particular pattern of meter and rhyme.

Morris's influence on nineteenth- and twentieth-century design was substantial. He focused mostly on textiles—wallpapers, fabrics, tapestries, carpeting, and the like—but he also worked in stained glass, typography, bookmaking, furniture design, and other domains of craft. His attitude toward these practices bears much in common with today's fascination with the authenticity of earnest, artisanal creation, and the rejection of the crude and impersonal capacities of industrial production. Morris was of a piece with today's Walmart-spurning, gastropub-frequenting design snobs.

For example: machine automation for creating fabrics and wallpapers was widely available in the 1860s, but Morris preferred traditional, manual methods of creation. "The noblest of the weaving arts," Morris writes, "is Tapestry, in which there is nothing mechanical: it may be looked upon as a mosaic of pieces of colour made up of dyed threads, and is capable of producing wall ornament of any degree of elaboration within the proper limits of duly considered decorative work."[27] This approach would take the name "Arts and Crafts," under which it became a thriving international design movement.

Morris's quip about art demanding resistance makes more sense when it's considered in the context of the Arts and Crafts aesthetic. Just as today's farm-to-table and local artisan movements are grounded in concerns for sustainability and environmentalism, so Arts and Crafts was a social and political movement as much as an artistic one. By focusing on the simple, traditional materials and methods for the creation of ordinary things like curtains and storefronts, proponents of Arts and Crafts believed that the pleasure and meaning of craft production would imbue ordinary life more broadly. If industrialization depersonalizes both the creator and the beneficiary of a craft object, then an Arts and Crafts attitude might offer a reminder of the connection between made objects and the conditions required to make them.

As a result, Arts and Crafts designs were often simple in form, created in a way that revealed the manner and materials of their creation. Morris's textile designs were naturalistic, often floral, with hand-painted details. While far more ornate than today's minimalist-design tastes would prefer, they were still far less artificial than machined patterns that had been popularized at the Great Exhibition of 1851. Likewise, Arts and Crafts architecture emphasizes the features of ordinary materials and construction methods, making brick, stone, wood, and tile decorative by highlighting their physical properties.

The twentieth-century Russian composer Igor Stravinsky had his own version of Morris's aphorism. "Human activity," he wrote in his 1959 book on the metaphysics of music, "must impose limits upon itself. The more art is controlled, limited, worked over, the more it is free."[28] If you find your eyes rolling a bit, I don't blame you. Proverbs like these work well for accomplished creators, because they rationalize the mysterious process of success after the fact. Just like hearing the appeal to the modesty of microhomes from the mouths of those who can afford the expensive materials and the property on which to build whatever they want, hearing the lesson of constraint from the mouths of acclaimed inventors feels like a retroactive rationalization. In reality, it's a long road from "resistance in the material" or "impose limits upon itself" to specific textile designs or musical compositions.

We must shift the frame from play as the creative exercise of human genius to play as the creation of playgrounds, where new materials become physically or conceptually circumscribed. The problem with most appeals to play, we discovered, can be found in their appeals to freedom and creativity. Play as the opposite of work, play as the antithesis of obligation, play as the Poppinsine magic that releases us from moil and duty and oppression. Play and creativity thus share much in common in the contemporary imagination. Both represent the exercise of freedom through supposed self-expression, meaning having been forged from within human hearts or minds or bodies and then loosed out of hands and mouths.

Despite its familiarity, the term "creativity" is actually a relatively new addition to the English language. The philosopher Alfred North Whitehead invented this neologism in his 1926 book *Religion in the Making* and refined it in his 1929 masterwork *Process and Reality*.[29] Perhaps surprisingly to modern eyes, Whitehead coined *creativity* for the philosophical discipline of metaphysics—the study of existence— not aesthetics or ethics or science or technology. And for Whitehead, the connection between creativity and existence lies in the problem of novelty. In his view, both philosophy and science had trouble accounting for how new things arise, whether ideas or mountains or recipes or action figures. *Creativity* was his invented word to name this process.

So far so good. Creativity sounds like what we thought: the inventiveness or vision or imagination that produces original works, whether by men or nature or God or accident. But our modern version gets things exactly backward. As the University of Essex sociologist Michael Halewood explains: "Creativity, according to Whitehead *is* a universal, in his very specific sense of the word, but this is not to imply that facts, things, or people can be placed within or draw upon creativity. Rather, it is the other way around."[30]

As Whitehead himself puts it, "creativity is always found under conditions."[31] Those conditions are much broader and deeper than human existence alone. Just as play names the conditions under which something can be manipulated, creativity names the conditions under which novelty can take place. Creativity always involves context, and not only the context of abstractions like interior design and background music and wealth and comfort. Whitehead is doing metaphysics, remember, not self-help or aesthetic theory or business consulting. Creativity is not a part of human experience, he urges, but a fundamental feature of existence. The fallacy of creativity, we might call it: mistaking our human exertion as the central factor in acts of creativity, rather than a peripheral one.

The fallacy of creativity helps explain why we don't really know what to do with the wisdom of William Morris, besides quoting it sagely and nodding as if he offers good advice. Even when creators talk about creativity in terms of constraint rather than imagination, something is still missing. Here's Stravinksy again: "The more constraints one imposes, the more one frees one's self of the chains that shackle the spirit."[32] Quotable for sure! But when faced with constraints—budgets, deadlines, conflicts, malaise, and all the rest—you'd be forgiven for failing to grok precisely how to free yourself from the chains that shackle the spirit. Constraint's creative potential quickly flips into a cruel Houdini escape trick that only geniuses can execute, while the rest of us drown. No wonder we embrace irony as an alternative, when the embrace of things is a stuntman's gambit. Constraint or no, creativity becomes a matter for thrill seekers with enough time, fortune, and nerve to be daredevils.

Charles and Ray Eames are the twentieth-century husband and wife designer duo who brought us the distinctive Eames lounge chair, the *Powers of Ten* film, and many other contributions to architecture, furniture, graphic design, and filmmaking. In a 1972 interview, Charles Eames offers a way out of the risky anxiety of creativity. After asserting that design depends largely on constraints, the interviewer puts the question to Eames, "What constraints?" He offers this reply:

> The sum of all constraints. Here is one of the few effective keys
> to the Design problem: the ability of the Designer to recognize as
> many of the constraints as possible; his willingness and enthusi-
> asm for working within these constraints. Constraints of price, of
> size, of strength, of balance, of surface, of time, and so forth. Each
> problem has its own peculiar list.[33]

Creative practice doesn't really arise from some unseen, divine inspiration that strikes an artist and that he or she subsequently carries out
in creative work. Rather, art emerges from a negotiation between a
creator, an initial vision or context, and a set of material limitations
that help lead the idea from abstraction to concreteness. Creativity is
always found under conditions. And so, if you seek the novelty that
births meaning, all you have to do is embrace the conditions rather
than mistrusting or rejecting them. You let them set the terms, but
you admit that *you* are also one of the terms. Circumscribing a set of
limitations always implicates you within it.

Then you have a choice. You can explore what is at play within
that playground, or you can reject the invitation and seek succor elsewhere. The latter option risks culturing ironoia, for it always holds
out hope for some better, more suitable situation, hoping dolefully
that it might eventually arise, and that if it did you'd even be able to
recognize it. The first choice is Charles Eames's: to embrace any constraint without prejudice, tackling any problem and any form and
any material. The Eameses' prolific and diverse portfolio of work is
testament to their promiscuity among constraint, their willingness
to accept almost any playground as the womb of a possible babe.

• • •

BUT THERE'S A third option, too, for not all of us are Eameses or
Stravinskys. Instead of acceptance or rejection, you can attempt to

fashion a new playground around a different set of materials, circumscribing another set of constraints and, in the process, discovering a
greater wealth of problems and solutions. Some of them you might
pass over dismissively. Others you might pursue enough to understand and respect. Others you might dive into feetfirst, or return to
after years away, or be cajoled into by a child or a spouse or a friend
or a parent. Like design, creativity reveals itself to have far less to do
with our own desires and visions and imaginations, and more to do
with the world outside us, and how seriously we are willing to take
it. How worldful we will allow our lives to become.

This latter and more modest version, by the way, is the one that
Stephie, the inventor of the phone-stacking game, prefers. "The basic premise is to just get people open to the idea of staying active and
attentive to one another," she wrote of the game on her Tumblr.
"It's really more of a fun concept in this new-age high-tech life of
ours." It's a rare, sophisticated answer to the problem of restraint
and constraint from anyone of any age. The phone-stack game isn't
a "life hack" or a "brain hack" or a clickbaity "this one weird trick"
or a clinically or even anecdotally psychologically effective mechanism for exercising willpower—it's just another thing you can do
with your phones at dinner. Sometimes you really want to talk, and
here's a way to ensure you dine with more chatter and less futzing
with your smartphone. But sometimes situations change, imposing
different constraints and changing the shape of the circumscription.
If someone needs to take an important call or gets a pang of anxiety about a forgotten task at work, you can redraw the playground
temporarily, admit these acts, and then reconvene the conversation.
Without Turkle's polite society "why don't we talk more" moralism.

This game isn't a radical idea or a life-changing realization. It's
a perspective, one that provides permission and tools with which to
free ourselves from the prison of restraint and to open our eyes and
arms to the infinite reconfigurations of stuff in the world that are

possible once we draw an imaginary circle around them and issue a commitment to them, even a tentative one, even a temporary one.

Today, we have the opportunity to restructure parts of our lives to focus more on constraint and less on restraint. Instead of seeing freedom as an escape from the chains of limitation, we need to interpret it as an opportunity to explore the implications of inherited or invented constraints. Luckily, we can find nearly endless examples of and invitations for constraint in the history of creativity, and in our contemporary lives as well.

SIX

The Pleasure of Limits

The history of creativity is a recurring story of adopting, accepting, inventing, and manipulating material constraints. By learning how others made playgrounds before us, we can better make our own.

CREATIVITY WASN'T ALWAYS THE GOLD STANDARD FOR value. As a decorative, political, or inspirational practice, art is a relatively new phenomenon. For most of human history, it was a tool for ritual and religious practice. Plato, for example, wouldn't have thought of painters or sculptors or musicians or dramatists as artists, but as imitators, creators of the ritual materials that allowed stories, forms, ideas, and myths to propagate.

For the Greeks, ideas didn't arise from within, but from without, via the Muses. These mythological figures embodied the source of knowledge and inspiration, which was transmitted through the worldly craft of the men and women who would carry out works of art at their bidding. Their creators would invoke the Muses when writing, as a ritual that called on these nymphs to speak "through" the author and also drew attention to the ultimate source of the subsequent material. You might recall the start of epic poems by Homer or Virgil, which commence with such incantations. "Sing to me of the man, Muse, the man of twists and turns driven time and again off course, once he had plundered the hallowed heights of Troy," begins the *Odyssey*.[1]

Poets were still called poets (*poïētēs*), but poetry meant something different for the Greeks than it does for us. Today, poets are artists who distill personal, internal feelings into literary form. But for the ancients, poetry was functional, not expressive. The word itself, *poiesis*, means "something made" (*poiéō* is the Greek word for "to make"). Poetry was a more celestial and communal project than it is today. It carried knowledge and history from the abstract past into the present and the future.

Already the Greek poet is relieved of some of the burdens of romantic creativity. Rather than expressing something crafted from within the individual self, the poet yokes a small and finite community to a long and immortal cultural tradition. Poetry becomes a worldly rather than a personal thing. The structure of culture over the accident of personality. And within that context, ancient poetry's form focused creators' attention to another structure, too: the form of the work, which is far more rigid than poetry today.

Ancient epic reads strangely to modern readers. We encounter it in codex form, bound between pages printed, cut, and glued into bindings (or perhaps digitized and delivered as files to handheld devices). These books are long, too, many hundreds of pages long, often published with flourishes like deckled edges, embossed jackets, or textured French flaps. These details are meant to make the work seem classical and refined, like a fine library tome from the late Enlightenment.

Except epics like the *Odyssey* aren't really books at all. Ancient epic wasn't written and read like novels and poetry are today, nor as they have been for the past millennium. Instead, epic was recited or "sung." Public performances of epic were necessary in preliterate society, but they also embodied an oral tradition that realized the nature of poetry as *poiesis*, as making. This making takes at least two forms. First, the making of the epic poem in the moment, through a reconstruction that creates the text anew rather than reciting it. And second, the making of a new connection to historical tradition, one important to the creed of the audience, through the reconstructed performance of the epic.

Epics are long, narrative poems. The *Odyssey* is over twelve thousand lines long, and the *Iliad* weighs in at more than fifteen thousand. If you've held the book rendition yourself, you know the heft and volume of thousands of lines of verse. The orator of an epic would hardly be able to memorize and recite the whole thing. So how did they do it? The answer illustrates the first of several key principles for managing limitations in the context of playgrounds.

LIMITS ARE MADE OF MATERIALS

Rather than memorizing an epic, a feat that would probably be impossible, orators made use of the material structure of epic poetry to help erect boundaries around the retelling of its story. Several material features of epic poetry helped make it possible to reconstruct a story like the *Iliad* from nothing.

First, epic wasn't exactly authored, not in the way even later narrative poems like Dante's *Inferno* were. We call Homer the "author" of the *Iliad* and *Odyssey*, and ancient historians did speak of the man as a real historical figure, even if they disagree on precisely when he lived. One convincing theory holds that Homer is the name of a particularly effective bard or orator, through which especially salient versions of the epics and hymns attributed to him entered the historical record. In any event, the origins of these poems were somewhat up in the air, even for the ancients. Combine the ambiguity of authorship with the appeal to the Muses and the role of the poet as a maker of an historical scenario rather than a timeless work of art, and the stage is set for a more flexible, interpretive performance of such a lengthy work.

Second, poetry was far more structured for the Greeks than it is for us today. The rise of free verse in the late nineteenth and twentieth centuries has made anything with unusual white space and line breaks qualify as poetry. But for most of literary history, poetry has conformed to specific structures. An English sonnet, for example, has a defined form, one that makes it a sonnet rather than a bunch of words on a page. Here is the first quatrain of Shakespeare's famous "Sonnet 18":

> Shall I compare thee to a summer's day?
> Thou art more lovely and more temperate:
> Rough winds do shake the darling buds of May,
> And summer's lease hath all too short a date; . . . [2]

Some of its structure is immediately apparent and familiar to the modern eye and ear. For example, the quatrain has an end-rhyme scheme: *abab*. Less obvious until you remember high-school English: he wrote in a particular rhythmic format, known as meter. Shakespearean meter is iambic pentameter, five sets of metrical feet with one short and one long syllable (each set is called an *iamb*): da DUM da DUM da DUM da DUM da DUM.

> ⏑ — ⏑ — ⏑ —⏑ — ⏑ —
> Shall I compare thee to a summer's day?

Put together, a sonnet in the Shakespearean style is fourteen lines of iambic pentameter, broken into three quatrains and a couplet, with the rhyme scheme *ababcdcdefefgg*.

For the Greeks, meter made language poetic. You'll hear no rhyming in Homeric epic, even in ancient Greek, but meter provided the structure that made the oral tradition possible. The meter of classical epic was dactylic hexameter, comprised of six metric units called *feet*, each of which is a *dactyl*—a long syllable followed by two short ones (although a dactyl could also be replaced by a *spondee*, two long syllables). "Syllable" is really the wrong word, though: ancient Greek wasn't like English, with a reliance on emphasis for meter. Instead, it was more like Chinese, with variations in tone and the length of different vowels providing rhythm and even a chant-like melody. This melody is why ancient epic was said to have been "sung" rather than recited. Here's the first line of the *Iliad*, first in Greek, then in Roman transliteration, and finally translated (in a fairly literal, unpoetic, and slightly misleading manner, so you can see what's going on).

μῆνιν ἄειδε, θεά, Πηληϊάδεω Ἀχιλῆος

Mēnin aeide thea Pēlēiadeō Achilēos

Of wrath sing, Muse—that of Peleus's son Achilles

If we split the line up, we can see its metrical units:

$$- \; \cup \; \cup \quad - \cup \quad \cup \; - / - \quad - \cup \cup \quad - \quad \cup \; \cup \quad - \cup$$

μῆνιν ἄ | ειδε, θε | ά, Πη | ληϊά | δεω Ἀχι | λῆος

That's two dactyls, a spondee, two more dactyls, and a special foot called a *trochee* (stressed-unstressed), which often appears at the end of a line.

Other material constraints apply to epic meter. A *caesura* or pause often appears in the second, third, or fourth foot. It falls in the middle of the third foot in the line above, right after *thea*, which means goddess (the Muse, in this context). And if the second or fourth feet are dactyls (long-short-short), the short syllables will typically be found in the same word (that's the case in foot four of the *Iliad*'s first line).

Okay, but what's the point of all this technical detail? To show that the epic poem was not only a story in the ordinary sense, but also a story created within the confines of specific constraints. Eventually the *Iliad* would be written down, but at first it was performed on the fly as oration, reconstructed from memory rather than recited word for word. The orator would have known the story of the Trojan War or of Odysseus's return home very well, but he would have had to extemporize the oration, a feat made easier by the required structure of epic meter. Much like a jazz musician improvises within the constraints of chord progressions, so the orator improvised within the constraints of passing metrical feet.

The bard had additional mnemonic tools at his disposal, too. Epics are filled with many human and natural characters, and many have nicknames to make them easy to remember. Known as *epithets*, they are rousing poetic atoms that hold up even without context: "the wine-dark sea," "Owl-eyed Athena," "Rosy-fingered dawn," "Swift-footed Achilles" (or alternately, "Achilles, son of Peleus," which we just saw in the first line of the *Iliad*). These epithets are easy to remember, and each also bears a predefined metrical format, ready to slot into a line of dactylic hexameter.

It would be an exaggeration to call the orator's work a snap-together affair, assembling tens of thousands of lines of epic like Legos or IKEA furniture. But the means with which to create the epic performance—the material structure of lines and feet and epithets and all the rest—are the scaffolds that make the epic what it is. Without them epic poetry wouldn't only be more difficult, it also wouldn't exist.

When William Morris talks about "resistance in the material," he puts the creator at center stage. You can picture her, striving to create or to find meaning but pressed back by the shadowy, Lovecraftian murk of *the material*. Flipping Morris's aphorism on its head offers a better take on how art (or meaning or identification or experience) really arises. Art doesn't take form *despite* material resistance, but *thanks* to it. Without a set of material structures and constraints, form itself would be impossible. A sonnet is not a poem *even though* it has to spill language across fourteen lines of iambic pentameter, but *because it does*. It takes the surplus of language and offers a rationale for structuring a portion of it in a particular way. Materiality isn't some cosmic surplus made into form through the enchanted genius of artists. It's what all of us do, every day, all the time. We may not be poets, but we are grocery shoppers and commuters, parents and marketing managers, knitters and amateur home brewers.

LIMITS CREATE POSSIBILITY SPACES

In 2008, the National Museum of Play in Rochester, New York, inducted three new toys into its National Toy Hall of Fame: the stick, the skateboard, and the baby doll. They were unusual choices, even by the standards of such an unusual palace of honor. More often, its inductees have been consumer goods—the Rubik's Cube, the Duncan Yo-Yo, the Atari 2600 video-game system, Barbie, Crayola Crayons, the Big Wheel tricycle, the board game Candy Land. But occasionally, common objects and abstractions have graced the plinths of this shrine: the blanket (2011), the cardboard box (2005), and the stick, the ultimate found-object-as-toy.

"It's very open-ended, all-natural, the perfect price," Strong curator of collections Christopher Bensch said of the stick, and who could argue.[3] He continued: "It can be a Wild West horse, a medieval knight's sword, a boat on a stream, or a slingshot with a rubber band. No snowman is complete without a couple of stick arms, and every campfire needs a stick for toasting marshmallows." But Bensch's litany of delightful applications of the stick followed a curious claim, although one nobody noticed at the time. "There aren't any rules for its use," he said. It's a sentiment that continued into the introductory paeans to the other two toys inducted that year. "It promotes imaginative play and brings out the nurturing side in all of us," associate curator Susan Asbury said of the baby doll. And by video conference, pro-skating legend Tony Hawk sang the praises of the skateboard, which he said promotes "individualism and artistic expression." Once again, creativity is mistakenly associated with freedom rather than limitation. To flip things around, we'll have to take a detour through the principles of design.

■■■

ONE ACCOUNT OF the role of constraints in design comes from Don Norman, a design scholar and consultant who first popularized the idea of "user experience" at Apple in the 1990s. In his widely read book *The Design of Everyday Things*, Norman suggests the concepts of "affordance" and "constraint" as tools for use in the design of user-centered objects of all kinds.[4] For Norman, affordances name the range of apparent possibilities that an object or device is capable of performing. A doorknob affords turning, a handle affords pulling, and a potato-chip clip affords clasping. Constraints, by contrast, limit apparently available actions. A square peg resists fitting into a round hole, the arrangement of connectors on a grounded electrical plug resists insertion in any way but the right one, and a heavy outdoor bench resists relocation by being too difficult to move.

Norman conceived his theory of affordances and constraints as a tool for designers to create apparatuses so that their users can easily and quickly determine the correct ways to put them to use. He famously ridicules certain designs of door handles, the ones that you try to push only to discover they pull instead, and the ones that fail to help you determine which side they hinge on, thanks to hiding the door mounts in the smoothness of uninterrupted glass on a building's facade. Once you know the term it's hard not to find these so-called "Norman doors" everywhere, especially as you fumble embarrassingly, bag or paperwork in hand, to enter commercial or government buildings that seem to have been designed expressly to suppress your ability to pierce their exteriors.[5]

Because Norman is concerned with the design of usable things, he focuses on the *apparent* affordances and constraints of objects. The door that looks like you can push it, whether you can or not, is a *perceived* rather than actual affordance. As such, the practice of user-centered design that affordance theory has championed entails creating alignment between apparent and actual use.

Key to this process is what Norman calls "natural mappings" between perceived and actual affordances—correspondences between what a thing looks like it can do and what it actually can do. One of Norman's unlikely bugaboos, for example, are ordinary light switches. The traditional switch maps "on" to the up position and "off" to the down one, but Norman finds great anguish in the disturbance of this convention in the case of two switches operating the same circuit—at the bottom and top of a staircase, for example, or on both sides of a room with two entrances. Norman devotes many pages of *The Design of Everyday Things* to proposing an enormous, ghastly design for a light-switch panel in which the position of the switches corresponds to the circuit that switch operates in the room—an earnest example that demonstrates the unseemly extremes of natural mapping.

Even though Norman's take on affordances is better known among designers, the psychologist J. J. Gibson first advanced the concept a decade before Norman's appropriation. For Gibson, affordances represent the action possibilities of a particular object.[6] While Norman's affordances are perceptual and psychological in nature, Gibson's are more promiscuous. *Any* potential or actual use of a thing counts, not only the ones that lead a specific human user toward success or failure for a given purpose, as in Norman's take. The heaviness of the outdoor bench resists tipping by wind as much as relocation by rogue park-goers.

If you think about Gibsonian affordances in relation to Heideggerian presence-at-hand, the sum of a thing's affordances represents all of a thing's possible interactions with all of the other things with which it might interact. All of its potential. A book can be read as a tool to disseminate knowledge. It can be displayed on a shelf as a social signal that identifies its owner as someone with particular interests and aspirations. For the right size book and the right size person, it can serve as a booster seat for a banquette stationed too low for its eager diner to maneuver over the table. It can act as a gift when handed over to another on the right occasion. It can become common property when stamped and labeled and put on duty in a library. It can act as a makeshift fan or a provisional visor on a hot, bright summer afternoon.

But also: a book can be rolled up and used as a deadly weapon against a fly or mosquito. It can be crammed under an open door to become a doorstop or under a toppling table to become a level. It can burn to produce heat (or censorship) when incinerated, yet it can also extinguish a candle's flame when closed hard against it. It can rot and fester and return to the earth when accidentally left behind at a campground. It can feed a goat.

■ ■ ■

SCIENTISTS AND MATHEMATICIANS sometimes talk about the "possibility space" of a problem.[7] Rather than making assumptions about a more or less likely answer to a hypothesis or behavior of an actor in an experiment, you could assemble, rank, and test all the possible directions a solution might take. In some cases, such as logical or mathematical puzzles, the overall solution space might be small, and a process of elimination can yield a result. The possibility space for a coin flip is heads or tails; of two coin flips, heads-heads, heads-tails, tails-heads, tails-tails. But for more complex problems, it's harder to draw a simple map of the solution space.

Possibility space is another name for the free movement made possible by play, a more technical term for the playground or magic circle. Once the material conditions of an object or scenario are clearly understood, the permutations of interactions—all the Gibsonian, real affordances the thing might allow—represent the perspectives or experiences possible in relation to it. Possibility spaces arise once a magic circle has been circumscribed, but before or between its actual use. While the ironist rejects a thing as insufficient and dangerous, the player uses its possibility space to ask questions about its potential. What can I do with the book? What can something else do with it? What does it do on its own? Or, what is possible to express in the form of the sonnet? What has already been done? What else can I do with its conventions?

If fun is the pleasure of finding something new in a familiar situation, then the fun of exploring a possibility space can entail proposing or discovering or executing some new angle on a familiar object, like the lowly stick-made-toy-hall-of-fame honoree. The limited possibility space of a doorknob might seem to make it a rather poor playground, until you need to hang a tote bag somewhere and the knob proves perfect, or its oval surface dents from use and you need to determine what brand and style it is to find a replacement, or it strikes you right in the small of the back causing brief, if acute, pain.

But sometimes the delight of exploring a possibility space comes from finding the very same thing once more, like the joy of siblings hurtling themselves outside to partake, yet again, in some real or invented game. Rediscovery also reinterprets Norman's purpose-oriented user-centered design. The doorknob works over and over again, day in and day out, over months and years and decades. It's a miracle worth savoring no less than playing a round of golf or fashioning a snowman in a surprise winter storm. That the universe could configure itself so that something would happen again just like it did before is no less novel than something never before seen.

Take sexual play as an example. We talk about "playing with" a partner, or a partner's parts, or a sex toy, but also of "playing with yourself." While this latter term is often a genteel way to say "masturbation," it hides a surprising insight about play and fun and limitations no less than it does about sexuality. Sexual self-play isn't only an indulgence or a taboo or a sublimation; it is also a means of discovering your body's proclivities and secrets for later exploitation, whether on your own or with friends. In this sense, "playing with yourself" refers both to the discovery of concealed features of your body—novelty in the ordinary sense—but also to the process of returning to familiar locations, moods, and physical manipulations. Sexual climax is perhaps the best and most obvious example of the pleasure of discovering the familiar again.

When it comes to Christopher Bensch and the Toy Hall of Fame, it's true that sticks don't have "rules" in the way that there are rules for an international soccer league or for Candy Land: dictates written down, imposed from without. A stick doesn't come with a manual. The local police or school board don't (yet) stop kids in the street for using sticks as horses or swords.

But that doesn't mean that the stick has no rules for use, as Bensch claims. A stick does have *properties*. Length and woodenness, strength and breakability, a status as detritus inviting its absconding and repurposing, sharpness and length, and so on. These features make possible all the uses Bensch outlines in his accolades of the stick—its capacity to simulate a saber, to become a cooking tool, to represent a snowman's arm, to be ridden like a beast or floated like a boat. The stick's delight owes a greater debt to the thing itself than it does to the child who would put it to use—even if that child's invention of new uses for the stick's material properties also contributes to its function as a plaything.

Among the misguided advocates of play-as-freedom, "rules" are often distrusted. Rules feel like structures of compliance, bureaucracy, control, and institutionalization. Rules impinge; rules dictate. And so, rules quickly become enemies of creativity, joy, and happiness. "Rules are made to be broken!" shout advocates of play like Sicart and Flanagan, right before they advocate for a different set of rules instead.

The difference between the stick as a device for supposed freedom from directed play and the stick as a thing that can be used in a variety of ways is also the difference between Norman's affordances and Gibson's. Norman gives us a designer's perspective, one that facilitates predictable and desirable outcomes. Norman asks, "What should I do with this stick?," to which the Museum of Play's Bensch responds gleefully, "Whatever you want!" If you'd like to test the profound failure of that directive, try it next time your kids or students or employees or church group asks, "What should we do?" By putting ourselves at the center of action, we set up our own anxious dissatisfaction. Instead, the answer is to circumscribe a playground, which is something parents (and teachers and bosses and pastors) do all the time. "Here's some paper and crayons; draw portraits of your grandparents," you tell the kids. "Write a one-thousand-word essay on the design of an object you use every day," you tell the students. "Find a way to reduce the unit cost of the product by 5 percent without adding a new vendor to the supply chain," you tell the employees.

Gibson offers a metaphysician's perspective instead of a psychologist's. His affordances ask, "What is possible?" And not only for you or for me, but also for anything else too—the otter or the gardener, the snowman or the forest or the ecosystem. Limitations create playgrounds. They help guide us, but they also remind us of the borders of the temporary domain we have made, borrowed, discovered, or erected. They are deliberate and visible, which means we can refer back to them and test our work against them. Is this portrait really of Grandma? Have I written enough, or too much, about my toothbrush? Have I considered all of the subcomponents that might be subject to efficiency or elimination to meet the cost target?

LITERACY AND VISIBILITY

I don't fault us for failing to understand the nature of epic poetry. Who could blame anyone for not knowing what they didn't learn? No matter what you might think about the Eurocentrism of the supposed canon of Western civilization, it's ironic that we'd know so little about epic, even though so many are compelled to read Homer in high school or in college as a part of a curriculum of time-tested Great Books.

And yet the constraints of the educational context draw a different magic circle that helps explain it. Few ever learned to read ancient Greek—only the aristocracy and the well-to-do, for whom tutors and boarding school were de rigueur, and today the aristocracy has largely given up on classics as a marker of the upper crust. Epic is oral, but nobody really knows for sure how Homer's Greek (itself a patois of different dialects) was pronounced. The best guesses we have come from the German and British classical traditions of the eighteenth and nineteenth centuries. Meanwhile, English (and French and Spanish and German . . .) don't lend themselves to dactylic hexameter and tonal reading like Greek does, and so many modern translators choose to fudge their efforts, rendering the epics in verse more familiar for their target readers, or even in prose. Today, literature isn't an experience of improvising and listening to a retelling, but of silently reading a book with serious contents.

It's certainly possible to appreciate the *Iliad* without any of the context of its historical form. But to do so, that appreciation must come in greater part from within us, in spite of the arbitrary and foreign nature of the work. We each have a certain tolerance for such directives, a patience for the magic circle a parent or a teacher or a curriculum or a culture draws around us, wagging its moralizing finger at us, rejoining us to "take this seriously."

We've even internalized this moralism, making it a feature rather than a flaw. Writing in the *New Yorker* on the ongoing debate about the value of an education in English in our technological era, Adam Gopnik has this to say:

> So why have English majors? Well, because many people like books. Most of those like to talk about them after they've read them, or while they're in the middle. Some people like to talk about them so much that they want to spend their lives talking about them to other people who like to listen. Some of us do this all summer on the beach, and others all winter in a classroom. One might call this a natural or inevitable consequence of literacy. And it's this living, irresistible, permanent interest in reading that supports English departments, and makes sense of English majors.[8]

On the one hand, Gopnik has the good sense to allow books to be books. They fill shelves and hands and libraries and hard disks, and they do different things in each of these cases. They are conceived and written, but they are also designed and printed and warehoused and wholesale price–negotiated and shared and displayed. He's taking books seriously. And for good reason: despite the lure of television, computers, and the Internet, more than three-quarters of Americans are reading books.[9] They're sometimes reading differently—on smartphones and websites, for example, but often not. A survey of children's reading habits by the publisher Scholastic found that 80 percent of the books kids ages six to seventeen read are the old-fashioned kind: paper glued between cardboard.[10] It's a good time to be a lover of language, even if it's also a difficult time to make a living at it.

But on the other hand, Gopnik divests bookishness from the world at large and reinvests it in the natures of individual human beings. I'm empathetic; I like books enough to have written a good number of them. But the idea that there are "book people" is no less insidious than the equally preposterous idea that there are "math people." Students of any subject tend to perform more poorly when they believe ability is more inborn and less malleable. This attitude becomes a self-fulfilling prophecy, since those students then take their supposedly lesser genetic predisposition as a justification for their own marginal performance.[11]

All of which is to say: we need not like, enjoy, or even be good at something in order to commune with it and to learn more about it. The mistaken idea that facility or enjoyment is a sign of adeptness—the "do what you love" mantra—mistakes fun for pleasure rather than exploration. But in order to explore the possibility space of a particular thing, no matter what it is, we must have sufficient knowledge about it to understand its limits, and then to deepen our response to that understanding.

■■■

CONSTRAINTS ON THEIR own aren't virtuous on account of being limitations. That view is the trap of restraint stated in different terms, like a passive-aggressive aunt urging that it's okay, she didn't want any coffee anyway, after you failed to buy the coffee you didn't know she wanted. In order to appreciate epic in relation to the constraints that facilitate it, we have to develop a literacy in the practice. That literacy isn't meant to produce greater knowledge or affinity for the thing in question, but simply greater understanding of and respect for its conditions.

The playground and the magic circle are metaphors, but they also can be taken literally. When you circumscribe a meadow and turn it into a soccer pitch, you do so by defining the edges of the field of play, by putting teams of players onto that field, by giving them a ball, by directing them not to touch it with their hands and arms, and so forth. Without these limitations, soccer would be incomprehensible. And indeed, if you watch an unfamiliar sport for the first time—maybe American football or cricket, depending on your nationality—your confusion mostly signals ignorance. In the face of ignorance, you can choose to reject an object as boring or otherwise insufficient, like the ironoiac does. But you always have another choice: to develop a literacy for that particular set of limits.

Don Norman and other advocates of user-centered design often proclaim the design value of *visibility*. If the actions that are possible to perform with an object are not visible, the user will fail to map them to desired outcomes. Visibility is the fundamental problem with Norman doors: if the door fails to make visible whether it affords pushing or pulling, or if it fails to make its hinge location visible so that the act of pushing or pulling can exert the rotational leverage necessary to open the door, then the user's effort will fail.

But user-centered design's idea of visibility asks nothing of the user. Objects are expected to fulfill preexisting needs without friction. That's fine when it comes to doors, perhaps, but it doesn't work for epic poetry or cooking or long-lingering August afternoons. In these cases, the ability to "evaluate the current state of the system," as Norman puts it, is not a subconscious psychological Gestalt—"Frontal cortex to motor cortex: push at your left, over"—but a messy process of voluntary attention.[12] And actually, why wouldn't we want the same to be true of doors, as well? After all, we encounter many more doors than we do poems anyway.

When creating or managing or redrawing the magic circles of ordinary experience, we want to cajole limitations into visibility, but not simply to coerce them into doing our bidding. Norman's model advocates for making a tool entirely ready-to-hand so that our brains can unconsciously do what the thing is supposed to do. Play encourages a more meandering route.

Think about the expert soccer player. She is constantly evaluating the conditions of the situation. The rapid, fluorescent hum of ironoia's sincerity or contempt has been replaced with brisk shifts of attention between the properties of body, pitch, ball, teammates, and opponents. The pleasure of playing arises from the constant pursuit of another perspective on what she already knows about the materials from which the game is made. Fun requires visibility, but not in the way that user-centered design does. The pleasure of limits arises no matter how easy or hard, no matter how automatic or arduous the process is of discovering something new about the system contained by its magic circle.

● ● ●

SOMETIMES THAT PLEASURE can be withheld until a new discovery reinvigorates your respect for an object. *Green Eggs and Ham*, the famous book by Theodore Geisel, aka Dr. Seuss, celebrated its fiftieth year of publication in 2010. Since the book is an icon of children's culture, its birthday was marked by news coverage it wouldn't have enjoyed otherwise.

Readers of these celebratory stories discovered something new about the familiar tome. The book didn't arise as the result of some special genius or hallucination of Geisel's, but as the outcome of a bet. Random House publisher Bennett Cerf challenged Geisel to write a children's book using only fifty words. The result would go on to sell over 200 million copies.[13]

The media published this story to mark the book's golden anniversary. And it's a good one; this feature of the book was new to me, even though it wasn't new to the world. Just like tennis had always contained Isner and Mahut's improbable 138-game tennis set, just as the Homeric epic has always been comprised of mnemonic-assisted dactylic hexameter, so *Green Eggs and Ham* has always contained but fifty words, arranged in arrangements strange and deranged. Knowing this fact deepens our respect for the book by showing us something new about it.

But Dr. Seuss's story goes even deeper: some of his other most famous works also arise from constraints. A 1954 *Life* magazine report on children's literacy suggested that kids weren't reading because children's books of the *Dick and Jane* variety common to that era were boring. After reading the study, Houghton Mifflin educational publisher William Ellsworth Spaulding assembled a list of 348 words that first graders ought to know. Spaulding asked Geisel, already an experienced children's book author by this time, to choose a subset of the list and to write a book with them.[14] Geisel managed to cram 238 of Spaulding's 348 words in *The Cat in the Hat*. According to some accounts, Geisel felt so constricted by Spaulding's list that he titled the book and selected its subject by picking the first two words on it that rhymed, *cat* and *hat*.[15] It didn't do so badly either, selling well over 10 million copies.

The truth of some of Dr. Seuss's most beloved works lies sandwiched between various constraints. Spaulding's word list, Cerf's even narrower one, and the narrowest one still, a book borne from any rhyme that might yield a plausible title.

The delight in finding something new in these familiar books doesn't take away from the enjoyment of their more affective experience: your less structured memories of their curious illustrations, their rhythm (it's anapestic tetrameter, by the way—da da DUM da da DUM da da DUM da da DUM), and the joy of their repetition.

Some might hold that pulling back the curtain on the creative process spoils it, ruins the magic. But the more visible the constraints of Dr. Seuss become, the more our appreciation of the work of art deepens. And in fact, there's reason to wish that people would have known these limits earlier and more broadly. The literacy-building drill function of *The Cat in the Hat* and *Green Eggs and Ham* might have encouraged more tolerance among parents subjected to otherwise insufferable repeated readings.

Limitations become more meaningful the more we know about them. A thing is not only what it appears to be, but it is also the conditions and situations that make it possible for it to be what it is. When it comes to constraints, the visibility we really want isn't one that helps us more easily "use" the things we take for granted, like doors or children's books, but one that helps us understand more deeply why they are the way they are.

LIMITATIONS IN CONTEXT

During the Great Depression, food was reasonably inexpensive, but families often didn't have the money to buy much. Guidebooks emerged for sharing recipes made from plentiful, cheap staples (cornmeal, rice, dried beans, etc.) that could be stretched across many meals. These practices got their start in even earlier guidebooks for frugal living. The 1832 handbook *The American Frugal Housewife* offers an assortment of tips and tricks, but some of its most delightful relate to stretching the use of foodstuffs. Roasted dry brown bread crusts are suggested as a substitute for coffee, or roasted rye grain soaked in rum, or roasted peas.[16] The book's author, credited on the title page only as Mrs. Child, admits that "none of these are very good" but that in a large family they may prove worthwhile, even if mixed half and half with coffee to stretch the contents.

I suppose we would call coffee made from peas or bread crusts "life hacks" nowadays, God help us, but Mrs. Child refers to them as "hints to persons of moderate fortunes."[17] A Calvinist delight in austerity pervades *The American Frugal Housewife*. At the time, the Protestant focus on work and frugality were not only exercises in restraint, but also acts of piety. It was important that the Joe taste bad in order that its drinker might imbibe the reverence associated with cutting the drink with peas, alongside the more familiar bitterness of coffee. One doesn't only go without; one also bathes in the gravity of such solemnity.

You can find a more ordinary version of Mrs. Child's deliberate asceticism in your refrigerator or pantry. We've all had to make do with less than we might choose, whether due to the financial exigency of unemployment or disability or because we didn't manage to make it to the store this week. In such a situation, the poverty of edibles becomes the circumscription that sets the stage for the meal. A gaggle of kids sent to the kitchen to "root, hog, or die," to use a phrase for fending for yourself from Mrs. Child's era, replaces the constraint of a healthful, balanced, prepared meal with the new playground of "anything goes" sustenance. At the risk of making light of the true wretchedness of poverty, even the meal served as a "best we can do" takes on greater urgency, every scrap bearing fathomless emotional value.

Pushed to its limits, wealth flips into poverty. In Silicon Valley, young, obsessive tech workers try to squeeze every last minute of coding out of their waking hours. Just as the late Apple CEO Steve Jobs wore the same uniform every day to avoid the distraction of choosing one's fashion, so software developers have begun to eliminate meals from their diets, replacing them with trendy protein powders like Soylent.

Swapping San Francisco's artisanal toast and cellophane noodles and bacon beignets with a loose, beige shake isn't for everyone, but those who choose it do so deliberately. "It just removes food completely from my morning equation up until about 7 p.m.," a twenty-three-year-old software engineer who goes by the situationally ironic name "Potluck" Mittal explained to the *New York Times*.[18] One need not endorse the replacement of the gastronomical and social delights of breaking bread with the lonely gurgle of chugging Soylent in order to appreciate that its adherents see the limitation as a deliberate attempt to increase their productivity. I'll admit that a Soylent diet seems perverse to me, a contemptible mockery of nutrition that only the wealthy could afford and the foolish could tolerate. But even so, I must also admit that Soylent's fanatics make clear that the fun of protein shakes is not directed at food, but at work instead. Remember: fun isn't maximizing pleasure, but finding something new in a familiar situation. Their situation isn't eating, but extreme labor.

Luckily, the well-(enough)-to-do have other options for culinary constraint. For the busy professional couple who didn't make it to Whole Foods, the preparation of food given what's on hand offers an opportunity to reconsider or temporarily subdue one's hatred for Swiss chard, or to seek new methods of preparation for the canned tuna and American cheese in the pantry and fridge. A website even assists with this exercise in constrained cuisine, supercook.com. Just enter the ingredients you have on hand, and the page updates with recipe suggestions, including any other ingredients you might need to rustle up to complete them. Should I find some ramen, Supercook helpfully advocates that I might fashion my canned tuna and American cheese into dorm-room cheesy tuna. Like Mrs. Child's pea coffee, most of the recipes look fairly ghastly, but the fun of a supercooked meal isn't really gastronomical anyway.

Entire television programs are now built around culinary constraints. *Iron Chef* was the first popular show of this sort, a Japanese import that eventually got an American rendition on Food Network. A world-class chef competes against one of the "iron chefs"

fictionally employed by a wealthy chairman, who has spent his fortune on an elaborate "kitchen stadium." A theme ingredient is revealed—sometimes ordinary, like cabbage; sometimes extraordinary,
like Kobe beef. The two chefs compete to prepare a meal of many
courses, for which a panel of judges marks scores.

But, alas, *Iron Chef* is something like the professional wrestling
of constrained cooking shows. The chefs are actually given consider-
able leeway, and the constraints impose fewer limits than appear to
the viewer. Off-screen, for example, they get time to plan their meals
after the secret ingredient is revealed, and only one of each dish must
be plated by the time the clock elapses. The theme ingredient is chosen with the competitors in mind, rather than acting as a diversion to
which they must adapt. While entertaining, *Iron Chef* is really a dra-
matization of constrained cooking rather than a performance of it.

For the real deal, you need to turn to *Chopped*, Food Network's
evolution of the competitive cooking show. The format is similar
to *Iron Chef* but even more demanding. In each of three rounds—
appetizer, entrée, and dessert—four contestants are given a basket
of three to five mystery ingredients. Each must prepare four plates
of a single dish in thirty minutes' time (twenty for the appetizer),
using *all* of the mystery ingredients, as well as any others they'd like
to draw from the show's well-stocked pantry. After each round, a
panel of judges appraises the results based on taste, plating, and use
of the secret ingredients. Based on their decision, one contestant is
eliminated, or "chopped," until a single victor remains after the dessert round. The contestants are drawn from ordinary working kitchens rather than celebrity ones, so skill, nerves, and familiarity with
the secret ingredients play a role in each chef's success. The winner
receives $10,000, a substantial windfall in the low-wage, high-stress
culinary industry.

The basket ingredients complicate matters. They are usually unworkable to the point of cruelty, often containing rare items that

will be unfamiliar to most, or processed foods, or proteins requiring time-consuming breakdown. A typical basket asks contestants to
make an appetizer from microwavable chocolate cake, baby tatsoi
(a Chinese cabbage), artichoke liqueur, and lamb heads. Or, an entrée from Vegemite, barramundi (an Asian sea bass), arugula, and
gummy snakes. Good luck.

Chopped is a joy to watch because somehow, against all odds, the
contestants always manage to get meals made, and delightful ones
too. Warm tatsoi salad with mole lamb head, for example—the chocolate cake inspired the mole sauce, a clever interpretation. Or, pan-
seared barramundi over a warm arugula salad with a gummy snake
beurre blanc. That the contestants are able to produce anything at all
is a miracle, but the food almost always looks and tastes (at least by
the judges' word) fantastic.

But *Chopped* has a problem. Unlike Mrs. Child's pea coffee or Potluck Mittal's beige protein slurry or the pantry Frankensteins you
might fashion with the help of Supercook, the show's judges assess
the results as if they were restaurant critics passing verdicts on
Michelin-starred establishments. It's easy to get tripped up. "I didn't
take the membrane off the tongue," chef Jordan Andino realizes after the appetizer clock elapses, looking down at his fried lamb tongue
over arugula with artichoke liqueur vinaigrette.[19] It's the kind of error the judges enjoy rebuking with a biting scorn that doesn't match
the playful and absurdist context of the game show. The producers
probably ask the judges to play down the lows and to play up the
highs for more dramatic television. But the results leave a bad taste
in the viewer's mouth, as if you were watching spent superheroes
being rebuffed for failing to arrive showered and dry-cleaned.

The feral absurdity of the basket ingredients—an outrageous
long shot from the viewer's perspective—becomes domesticated
once the judges get their forks on the resulting plates. "The chocolate sauce still tastes like chocolate cake mix," judge Amanda Freitag

complains to chef Ingrid Wright, concocter of the mole lamb head. "We were really looking for you to transform that."[20] (Wright gets chopped that round, alas.) Chris Santos, a *Chopped* judge and executive chef at two New York eateries he also owns, expresses brow-furrowed reproach at a contestant, urging that food amalgamated under the show's duress also somehow be "something you would serve in your restaurant."

Constraints suggest a context. They set the terms for the experiences that take place within them. By making the limitations at work in *Chopped* unambiguously visible to the contestants, judges, and viewers, the show sets up the expectation that the lava-filled moat that comprises its extreme magic circle also encircles the judges on their plinth. But more often, the members of the jury seem like spoilsports or bullies intent on changing the rules to accommodate their shifting preferences. For Johan Huizinga, the spoilsport is an agent who refuses to play the game and, in so doing, "shatters the play-world itself," robbing it of its illusion.[21] The pleasure of limits arises only when the participants within a particular magic circle understand and respect the material constraints it circumscribes. Otherwise, they are playing with a different object entirely, mistaking peas for coffee rather than embracing the monstrosity that is coffee made from peas.

■ ■ ■

CONSTRAINTS ARE MOST effective when those who are bound up with them can clearly see, understand, and appreciate the limits they impose. That doesn't necessarily mean *accepting* those limitations as a best approach to a pursuit, nor does it mean fixing them for eternity as the only way to do things. Rather, it entails creating a playground in which the limits, materials, participants, and context are clear and visible.

Compare *Chopped*'s version of constrained competitive cooking to a more effective but equally power-skewed situation: professional pitches and presentations. If you work in or around sales of any kind, you may be familiar with the "elevator pitch." It's a succinct, clearly expressed summary of a product, your resume, your charity, a business proposal, etc. It's called an elevator pitch because it's supposed to be short enough to deliver during an elevator ride. But even if it's hawked in a different context, other features are implied. Elevator pitches are conducted between interlocutors of uneven station, for one. And for another, they are often unexpected. Those who perform them may have fallen upon the attention of a more important person unexpectedly, whether in the elevator or the lobby or the golf club or the gastropub. The elevator pitch is a cold call, a first encounter that succeeds if it leads to a later, more in-depth encounter.

It's a tough genre. Elevator pitches demand improvisation and gregariousness, yet they also underscore inequalities of access, wealth, and power. But the elevator pitch also creates a domain for mutual understanding that momentarily suspends some of those imbalances. And when deployed abstractly, it forces its performers to develop a short, clear, and direct explanation of their offering that anyone can understand.

The elevator pitch is normally thought of as a business practice, but its lessons apply more broadly. Increasingly, clear and visible constraints have become a part of creative practice more generally. Since 2001, the 48 Hour Film Project has challenged filmmakers to create a short film within two days' time, given the filmic materials similar to the secret ingredients on *Chopped*: a genre, a character, a prop, and a line of dialogue. Similar contests have developed in other media forms, too. The video-game version of 48 Hour Film Project is the Global Game Jam (GGJ). Over a weekend, game developers around the world make a game given a theme assigned to their time zone. When audiences engage with the words that result from these gatherings, they attenuate their expectations to take into account the conditions under which they were made.

More constraint isn't always better constraint; sometimes just enough is best. November is National Novel Writing Month. Its aim: to get writers to compose a fifty-thousand-word novel in thirty days. While anyone can participate, NaNoWriMo (as its participants affectionately call it) is particularly appealing to first-time or would-be novelists who have always wanted to try their hand at novel writing but have always fallen prey to excuses or roadblocks.

The quality of the novel (or film or game) resulting from a constrained creative context might be subpar, but quality is less important to NaNoWriMo than completion. As writer Joel Cunningham puts it: "just getting it done is a challenge, never mind actually coming out at the end of it with something readable."[22] But like the elevator pitch, GGJ and NaNoWriMo succeed because they set their expectations clearly: the completion of a first go that might lead to more later, in the context of a community who have all agreed to support that circumscription of the creative process. And sometimes it does. Erin Morgenstern's 2011 novel *The Night Circus* started as "fifty-thousand words of unconnected scenes and imagery," in Cunningham's assessment, during NaNoWriMo 2004. Seven years and many more months of writing and revising later, the book was a #1 bestseller.

LIMITS ARE BOTH CONVENTIONAL AND INVENTIVE

You might or might not enjoy lamb head or Soylent, but like everyone, you have predilections. Maybe you love romantic comedies or detective novels or soft jazz albums or reality-television shows or meaty wild mushrooms. How do you know that something in particular is the thing you enjoy? Sometimes it tells you directly, on the cable guide or the chalkboard menu or the bookstore shelf. But in other cases, you intuitively recognize the things you prefer and separate them from the ones you don't. Genres and forms offer good examples of the thrill of rediscovering the familiar, but also of the way pleasures are associated with ground as much as with figure. Often we believe that delight *must* surprise us, that the novelty of fun appears dramatically against a high-contrast backdrop. But even more frequently, pleasures sneak up on us when we're not watching. The mere recognition, even unconsciously, that a thing is *your kind of thing.*

On further analysis, the patterns that found our preferences have material structures. And experts, scholars, snobs, and fans pursue the depth of knowledge and practice reserved for zealots. But zealotry is precious, and any individual can only hope to have one or two deep devotions at a time. Whether or not expertise takes the 10,000 hours of practice that Malcolm Gladwell famously argues it does, expertise always melds clarity with blindness.[23] Like fandom, expertise is obsessional, and obsession narrows our field of view, filling our time and attention with one subject in depth, where a thousand could fit superficially.

But depth also teaches us about this blindness, in at least two ways. First, by drawing attention to our obsessions against the backdrop of the missed opportunities they preclude, and second, by teaching us what it feels like to go deep in the first place. Whether it's chess or cheese making, knitting or nightclubbing, the experience of expertise is transferrable. Once you've devoted yourself to any single commitment, you've exercised the muscle sufficiently to bring that respect to others.

The aerobics for this sort of physical education comes from learning to recognize convention. For some things, even minimal encounter entails an understanding of convention. The sonnet offers a good example: knowing that you're even looking at a Shakespearean sonnet is more or less coextensive with knowing that such a sonnet is fourteen lines of iambic pentameter across twelve lines of individual, alternating end-rhymes and a couplet. Should you ever try to write a Shakespearean sonnet, you know how to get started. But in many cases, even very familiar ones, the materiality of a favorite thing isn't obvious. You can easily imagine receiving a gift of Shakespearean sonnets and finding them delightful without knowing how they are materially constructed.

In fact, this happens to us all the time. Structures are often hidden from view. Most of us listen to the Police or Madonna or Taylor Swift on the radio during our commutes, but how many know that the standard structure of a pop song is verse-chorus-verse-chorus-bridge-chorus? (Or what a musical bridge is? It's a contrasting section that prepares the song for a return to the chorus.) Most of us watch *Everybody Loves Raymond* or *Cheers* or *The Simpsons*, but how many know that a sitcom also has a common pattern: beginning, plot point one, commercial break, middle, plot point two, commercial break, resolution. Screenplays have standard structures, of which three-act, twelve-act, and nine-act are the most common. Foods have structure too. Ice cream is made with cream and whipped fast, making it higher-fat but lighter and more voluminous, while gelato is made with milk and churned slowly, resulting in a denser, lower-fat product.

Conventions arise over time and through tradition. The constraints of the oral tradition—mnemonics, meter, and the like—are evolutionary. The epic arose in mysterious ways over many centuries of human civilization, most subject to unrecorded or lost history. The sonnet, by contrast, was invented and refined by a series of early Renaissance writers, including Petrarch, the poet the Italian Petrarchan sonnet was named after. When new forms like the sonnet inspire an eruption of creative and receptive energy, we bathe in the novelty of their surprise. Eventually, if fortunate, figure becomes ground and the unfamiliar becomes commonplace. Success begets disappearance, and the most prevalent structures become the water in which we proverbially swim. A medium is successful when you don't even notice it anymore. Ground is always background, and we don't notice it until we have new reason to do so. Sonnets have long since ceased to be figure, and even as ground, we mostly think about them as "some kind of poem we had to read in school once" rather than as an active part of our world willing to accept new attention.

It's not necessary that you put down this book and start reading sonnets again, let alone writing them. But if you did, you'd find all sorts of fun awaiting you. Petrarchan-sonnet fan fiction about Jem and the Holograms, perhaps, or Shakespearean-sonnet renditions of the e-mail exchanges from your neighborhood listserv or your human resources department. But even to ponder such a return to the long-grounded sonnet, you'd have to practice. It wouldn't come for free, and you'd have to submit yourself to the sonnet's form, willingly and without compromise.

Accepting limitations helps us appreciate the relationship between the constraints that underwrite a particular thing and all the other magic circles we might have circumscribed instead, whether around similar or different subjects. To transform the particular limitations that constitute a convention from ground into figure—whether film genre, casserole recipe, or landscaping technique—we must practice managing limitations as figures in the first place. One way to do this is through invention rather than convention.

■ ■ ■

A MATHEMATICALLY ORIENTED collective of writers known as the Oulipo has been doing such work since the mid-twentieth century. Oulipo—short for *Ouvroir de littérature potentielle*, or "workshop for potential literature"—adopts or invents writing constraints and uses those constraints as the foundation for new writing.

You're probably already familiar with some of the constraints that Oulippians have recuperated. One is the *palindrome*, a word or phrase that reads the same backward and forward. "A man, a plan, a canal Panama" is a famous palindromic one-liner about the conception of the Panama Canal, and likewise, "Able was I, ere I saw Elba," a somewhat obtuse reference to Napoleon Bonaparte's exile to the Mediterranean island of Elba in 1814.

These quips may feel like groan-worthy dad jokes, but some historical precedent demonstrates their role as an exercise for and expression of a willingness to look at familiar problems in a different light. The group of mathematicians and cryptographers who worked as code breakers at Bletchley Park during World War II, for example, had a penchant for palindromes. Alan Turing is probably the most famous of these figures, thanks not only to his role in breaking the cypher of the Nazi Enigma machine, but also for his later invention of the architecture of the modern computer, not to mention his recent introduction to a mass market in the Academy Award–winning film *The Imitation Game*. But a lesser-known British mathematician, Peter John Hilton, was apparently the better palindromist among WWII code breakers. He is credited with authoring this gem of a palindrome during Bletchley Park's unofficial competitions: "Doc, note: I dissent. A fast never prevents a fatness. I diet on cod." At fourteen words and fifty-one characters (not including spaces and punctuation), it's an impressive feat that delights, despite making dubious sense.

But these one-liner palindromes look like amateur hour next to the work of the Oulipo. In 1969, Georges Perec wrote "Le Grand Palindrome" ("The Great Palindrome"), an attempt to treat the palindrome as a legitimate literary form, akin to the sonnet in literary station. "Le Grand Palindrome" is 5,566 letters long, or about one thousand words. It takes the form of two texts split at the center, mirroring one another.[24] The result is cryptic at best, and many readers will find it baffling. Here is a portion of the start and end of the work, both in French and as translated (unpalindromically, alas) by the Ouilippian Harry Mathews:

> Trace l'inégal palindrome. Neige. Bagatelle, dira Hercule. Le brut repentir, cet écrit né Perec. L'arc lu pèse trop, lis à vice-versa. . . .
>
>   . . .
>
> Désire ce trépas rêvé : Ci va ! S'il porte, sépulcral, ce repentir, cet écrit ne perturbe le lucre : Haridelle, ta gabegie ne mord ni la plage ni l'écart.

> Trace the unequal palindrome. Snow. A trifle, Hercules would say. Rough penitence, this writing born as Perec. The read arch is too heavy: read vice-versa. . . .
>
>   . . .
>
> Desire this dreamed-of decease: Here goes! If he carries, entombed, this penitence, this writing will disturb no lucre: Old witch, your treachery will bite into neither the shore nor the space between.[25]

Under the radar, the dream of the literary palindrome persists. In 2002, Nick Montfort and William Gillespie published a 2,002-word long palindromic story called *2002*, which is currently thought to be the longest known palindrome. While stylistically unusual from the perspective of the ordinary short story ("Bob's eyes, baby octopi, led her on. Bob steps. Babs, anon. Rob all"), Montfort and Gillespie succeed at producing a work whose unusual form is inextricably bound up with its contents.[26] Both resist becoming ground, and the serious reader will be forced to gestalt rapidly between the plot, curious as it is, and the form that produces it.

The clear relationship of form and content is a central feature of invented constraint. Some Oulippian constraints, like the *lipogram*, adopt a more subtle balance between the two. A lipogram is a text that omits one or more letters from its composition. It is an ancient form; Nestor, a poet of the third century BC, even wrote a lipogrammatic version of the *Iliad*. The epic is divided into twenty-four books, a number that corresponds conveniently with the quantity of letters in the Greek alphabet. Nestor's rendition omitted alpha from the first book, beta from the second, gamma from the third, and so on.

Perec composed one of the most famous lipogrammatic works, a novel called *La Disparition* ("The Disappearance"), which is a lipogram in *e*—that is, an entire book in which the commonest letter in the French language is omitted.[27] The coupling between the form and content is deliberate. The book is about a group of friends searching for a missing hunting companion, who also discover, bit by bit, that another more obvious symbol is also missing. The missing *e* becomes a symbol of loss more generally, and the work can be read as an elaborate, unexpected formal meditation on coming to terms with loss. The book was translated into English in 1995 by the novelist Gilbert Adair as *A Void* (there's an *e* in "disappearance," unfortunately)—perhaps an even greater accomplishment than Perec's original.[28]

You can take things much further, if you'd like. One of the Oulipo's invented limitations, the prisoner's constraint, is a lipogram in ascenders and descenders—that is, a work in which there are no letters with portions of their forms extending above or below the x-height (the height of a typical minuscule letter). Next time you find yourself bored at the office, try composing your memos or e-mails without *b, d, f, g, h, j, k, l, p, q, t,* or *y*.

Occasionally, making constraints visible also leads to novel inventions that become conventionalized. 48 Hour Films and NaNoWriMo novels sometimes become works in their own right, but the form of the two-day short film or the month-long novel is mostly a means to an end judged in relation to its limits. But PechaKucha, a constrained presentation format first begun in Japan, has made the leap from tentative magic circle to full-fledged public-speaking genre.

First conceived in 2003 as a way for designers, particularly architects, to share work in their communities, the format is based on the traditional PowerPoint-type slide show. But instead of endless, lingering screens of bullets and lists, PechaKucha imposes a strict format: twenty slides, each displayed for twenty seconds before automatically advancing. The result is a fast-paced, six minutes and forty seconds long overview of an idea or set of works. Eight talks can fit in an hour once accounting for quick transitions, making PechaKucha a great format for a dynamic, short evening of sharing that still leaves time to socialize and further discuss ideas.

Like the now-hallowed eighteen-minute TED talk, the official PechaKucha name and its associated event, a PechaKucha Night or PKN, are trademarked and closely controlled (by Klein Dytham Architecture, who invented the format). But that doesn't stop anyone from adopting it, quickly explaining it to their audience (often done as the first talk for unofficial gatherings), and setting the machine in motion. PechaKucha has found its way into all varieties of business, educational, and creative contexts.

In addition to its formal, visible constraints, the format also imposes a more informal, implicit one. A series of images each displayed for twenty seconds works well for designers to show a portfolio of their work, the original purpose of the format. But it's harder to make a formal argument in the PechaKucha format. When used in professional and academic conferences that share ideas or abstractions more than specific works, some organizers invoke different names and constraints, like "lightning talks" with five-minute timeframes but absent auto-advancing slides.

No matter the specifics, the result is a family of genres of public speech, all of which use limits to force the condensation of ideas into performances—part elevator pitch, part PowerPoint, and part stump speech. On first blush, the PechaKucha seems like it would be easier to compose and perform than the hour-long formal lecture. But like the elevator pitch, it's actually much harder to condense ideas into a succinct but intelligible format. This format doesn't result in some kind of depraved, Taylorist improvement in the efficiency of conferences—more ideas per hour!—but a refinement of the quality and digestibility of those ideas.

COMPUTATIONAL CONSTRAINT

Derived from PowerPoint, PechaKucha also shows us how the digital age undergirds opportunities and materials for drawing new magic circles around familiar computational systems. Technology neither frees us from all limits thanks to digitization nor imprisons us in the nefarious trances of screens. Instead, computation has made invented constraint a familiar part of our everyday lives.

In the early heyday of the personal computer, ordinary computer users enjoyed more frequent encounters with the limitations of technology than we do today. Files had to be saved on cassettes or floppy disks. Even when affordable, hard disks were still profoundly small, making file management a regular concern. RAM was limited even for common software usage scenarios, and computer owners had to determine if programs they bought at the store would run on the hardware they had at home. Even earlier, in the mainframe and minicomputer eras of the 1960s and '70s, computer users had to schedule time to feed punch cards or run programs on time-sharing systems.

Today, we think of technological progress as the ongoing, endless advancement of capacity. Moore's law predicts the doubling of transistors on microcontrollers every eighteen months. New devices and services are sold as means to achieve greater and greater power and freedom, whether by means of being able to render ever-more realistic 3D environments with a new graphics card and game engine, or by being able to outsource our chores to newly found underlings. To be sure, the creators of these systems face constraints all the time—but technical constraints are usually framed as hurdles to be overcome through cleverness and progress and grit and overtime.

But even as we careen toward an increasingly technophilic future, some contort the computer into a medium for invented limitation. Since the first popular microcomputers of the 1980s standardized hardware and made computer graphics and sound affordable, participants in the "demoscene" have held parties and competitions to show off as much computer prowess as can be accomplished in as little code as possible.

A characteristic example is the PC demo Elevated, created in 2009.[29] At first it looks like the kind of 3D real-time graphics show we have become accustomed to, thanks to video games and computer graphics (CG) filmmaking. You appear to be flying over a realistic mountainscape, covered in ice and snow and rock and water, the sun's light refracting from clouds above. Electronic music plays, synchronized with the motion of clouds and light. As a CG demo, it might have impressed someone in the early 1990s, but these days even middle-grade direct-to-video movies have effects this good.

But not in four kilobytes (KB). That's the total size of Elevated's executable. It's able to be so small because it doesn't contain any complex 3D graphics or animation, any textures or compressed music files or animation rigging. Everything is created on the fly as the program runs, from code—the geometry of the mountains, the lighting effects on the water and in the air, the music that accompanies the visuals. For comparison, the three-minute, medium-quality AVI movie of the demo I downloaded to watch it sucks up 67.5 megabytes (MB) of space, more than 17,000 times the size of the demo that produces the movie that file stores. Jaw dropping, but only in the right context. While mostly awestruck, some of the scenesters who post and comment at the demoscene site pouet.net were still nonplussed. "Very nice landscapes, but the music is so-so," wrote one, before concluding, "Still, definitely a thumb up."

...

AN EVEN MORE constrained, more esoteric, and more Oulippian version of computational invention than the demoscene can be found among those who create computer languages. Most ordinary computer languages are created for facility, power, purpose, and the audience of programmers to whom they cater. They are typically generic, geared toward specific computational approaches rather than particular genres or applications of software. C and C++, for example, offer low-level control and speed. Java offers portability to many platforms. And Python offers a simplified syntax and high-level control that make it possible to write programs that would need to be far more complex in other languages.

But just as Georges Perec and his ilk lurk in the alleys of literature, so a similar crew of computational inventors have devised entirely novel computer languages whose purpose is to draw attention to the diversity of expression possible in a format that is normally considered entirely mechanical, driven by efficiency and purpose.

The programming language C was first developed in the early 1970s, and it has remained a popular general-purpose, low-level tool for software development on many platforms. It is a flexible and challenging language with a reputation for ambiguity and syntactical confusion. In ordinary production code, software engineers seek to reduce this turmoil in order to produce readable and maintainable code. Computer programmers generally embrace the aesthetic values of simplicity, legibility, and efficiency.

But since 1984, the International Obfuscated C Code Contest (IOCCC) has asked its entrants to turn those values on their head. Instead of writing elegant code that disambiguates, IOCCC programmers reinterpret the language as a different sort of playground. Each entry is invited to harness the unusual features of C in order to write the most confused, perverse C program possible.

Among the many entries and winners in the contest's thirty-plus-year history, many require knowledge so esoteric that it would be onerous to try to explain them. The creative use of white space is common (the C compiler ignores spaces, line breaks, tabs, and the like), and many entries look like concrete poems, their instructions and variables and preprocessor directives taking on the physical form of the program that code executes. But one far more sophisticated entry can be enjoyed by anyone, whether or not you've ever programmed a computer. Here's how the code from Brian Westley's winning 1990 entry begins:[30]

```
char*lie;
    double time, me= !0XFACE,
    not; int rested, get, out;
    main(ly, die) char ly, **die ;{
      signed char lotte,

dear; (char)lotte—;
    for(get= !me;; not){
    1-out & out ;lie;{
    char lotte, my= dear,
    **let= !!me *!not+ ++die;
      (char*)(lie=
```

They are clearly letters between two disgruntled lovers, Charlie and Charlotte. It takes a bit of squinting to read them as natural language amidst the mess of unfamiliar C symbols. But as literature, Westley's program is at least as readable as Perec's or Montforts and Gillespie's palindromic poems.

The program takes advantage of C's keywords and syntax to make the letters between Charlie and Charlotte also compile as code. The first line, for example, declares an alphanumeric variable (char* is a pointer to an array of characters, or letters) called "lie."

Things don't end well for Charlie and Charlotte, unfortunately ("get-!out; if (not—) goto hell;"), but when the program is compiled and run, it does something delightful.

There's an old ritual called the daisy oracle, which is recited while plucking the petals from a flower. The couplets "he (or she) loves me," "he (or she) loves me not" are uttered in sequence with each pluck until all the petals have been removed. The line you speak when plucking the final petal is supposed to reveal your beloved's fated feelings toward you. Westley's program works as a daisy-oracle simulation. When you run it, you tell it the number of petals in your flower, and it runs through the process, reporting back either "loves me" or "loves me not."

Individually, neither the strangely formatted correspondence of the source code nor the computationally superfluous operation of the program (let alone its matter-of-fact output) is particularly inspiring. But just as contestants on *Chopped* delight us by producing viable dishes given preposterous ingredients, so Westley's program offers an unexpectedly dense little truffle of pleasure when experienced on all its registers. It shows that it is possible to do so many things all at once, and in such little space: to make a computer program about love that is *also* an experimental poem about love.

In an article about the aesthetics of computer programming, Michael Mateas and Nick Montfort call this technique "multiple coding."[31] A multicoded work is legible and meaningful on multiple

registers all at once: as a sharp commentary on the ambiguities of
C syntax; as a tiny fiction; as a daisy oracle; as a fortune-teller for
romance. And moreover, each of these registers relates to the others
thematically, tying the result up in a bow. Admittedly, the first is a little hard to understand outside the world of computer programming,
but, suffice it to say that C is a difficult language to love and a difficult
language to leave—and on top of it, computer programming hasn't
traditionally been associated with romantic or sexual prowess (or,
as some compilers output when processing Westley's source code,
"warning: eroticism unused in function main").[32]

The Australian astrophysicist David Morgan-Mar takes multicod-
ing even further. Instead of exploring the play intrinsic to a widely
used language like C, he designs entirely new "esoteric programming
languages" that bring the multilevel semantics of Westley's daisy
oracle to a general-purpose instruction set that can be used to produce any number of new programs.[33] Among his creations is Piet,
a visual programming language that uses image files as input. The
program reads the image from the upper left corner, and executes
instructions that correspond with changes in the hue and lightness
of the color it encounters. It's called Piet because the programmer
can create source code that resembles the geometric abstract art of
Piet Mondrian.

But my favorite of Morgan-Mar's unholy esoteric languages is
Chef, a language in which the programs are also recipes. It's less crazy
than it sounds; after all, programming and cooking are similar: both
execute step-by-step processes on materials, digital or culinary. Computer programs have memory, instructions, input, and output. The
memory is stored in variables, the input is taken from the keyboard
or mouse or disk drive or network, and the output is directed to the
screen or speaker. In Chef, the variables are "ingredients"; the input
is the "refrigerator", the output and memory is a series of "mixing

bowls" and "baking dishes," and the instructions are the steps of the recipe itself.

For example, you might tell Chef that your program needs six diced onions, in which case the value "six" would be assigned to a variable named "diced onions." Later, you might tell the program to "put diced onions into the mixing bowl." This would place the value of "diced onions" (i.e., six) onto a data structure called a *stack*, which is represented by the mixing bowl. All of the language's mathematical and stack-manipulation operations have double meanings that relate both to computation and to cooking: add adds; remove subtracts; divide divides; stir rolls the stack, and so forth. Here's an example, the canonical "hello world!" program that every programmer writes first when learning a new language.

HELLO WORLD SOUFFLE.
This recipe prints the immortal words "hello world!" in a basically brute force way. It also makes a lot of food for one person.

INGREDIENTS.
72 g haricot beans
101 eggs
108 g lard
111 cups oil
32 zucchinis
119 ml water
114 g red salmon
100 g Dijon mustard
33 potatoes

METHOD.
Put potatoes into the mixing bowl. Put Dijon mustard into the mixing bowl. Put lard into the mixing bowl. Put red salmon into the mixing bowl. Put oil into the mixing bowl. Put water into the mixing bowl. Put zucchinis into the mixing bowl. Put oil into the mixing bowl. Put

lard into the mixing bowl. Put eggs into the mixing bowl. Put haricot beans into the mixing bowl. Liquefy contents of the mixing bowl. Pour contents of the mixing bowl into the baking dish.

Serves 1.

I've been making my Georgia Tech students write Chef programs for more than a decade, so I'm probably responsible for coercing 99 percent of the world's Chef code (and cookery) into existence. But when I conduct this assignment, it's not enough only to write a Chef program that does something. The best submissions are multicoded, operating on many registers all at once. Each layer adds more difficulty, depth, play, and fun to the overall specimen:

1. A program that reads like a real recipe, not just a weird amalgam of recipe-language
2. A program that runs and does something vaguely interesting when executed on the computer
3. A program whose execution relates to the recipe thematically
4. A recipe that can be followed logically in the kitchen; one that doesn't contradict itself
5. A recipe that, when cooked, produces actual food rather than just culinary rubble
6. A recipe whose culinary output is gastronomically pleasing

This list is more difficult than it looks. If you look back at the "hello world" recipe, you'll spot some curiosities. One hundred and one eggs? Thirty-two zucchinis? These numbers are so large because the program uses the Unicode character values to produce text (for example, 101 is *e* and 32 is a space). And so the clever Chef multicoder has to perform acrobatics upon the recipe to make it work as both a cooking and a computational experience. In some cases, unnecessary or repetitive kitchen acts are required in order to make the dish also work as a program.

Here's a successfully multicoded Chef program, written by one of my former students. (After cutting his teeth on Chef, he went on to work on commercial video games like Uncharted.)

IRISH CREAM DESSERT SQUARES.
This recipe creates sapid dessert squares perfect for party pleasure, and it also displays the square of any number.

INGREDIENTS.
1 package of yellow cake mix
3 beaten eggs
12 tablespoons Irish Cream Liqueur
5 tablespoons oil
1 can of cream cheese vanilla frosting
61 white chocolate chips
Cooking time: 30 minutes.
Preheat oven to 180 degrees Celsius gas mark 4.

METHOD.
Take beaten eggs from refrigerator. Put beaten eggs into mixing bowl. Combine package of yellow cake mix into mixing bowl. Combine beaten eggs. Mix the mixing bowl well. Put oil into mixing bowl. Fold Irish Cream Liqueur into mixing bowl. Mix the mixing bowl well. Pour contents of the mixing bowl into the baking dish. Put can of cream cheese vanilla frosting into 2nd mixing bowl. Combine white chocolate chips into 2nd mixing bowl. Liquefy contents of the 2nd mixing bowl. Pour contents of the 2nd mixing bowl into the baking dish.

Serves 15.

This recipe does something computationally simple—calculates the square of an input number—while simultaneously baking dessert squares (see what he did there?). I always invite the students to cook and bring in their Chef homework, and I can assure you that John's Irish cream dessert squares taste delicious too (go ahead, try it out yourself).

PUSHED TO THE LIMITS

This is great, but what's the point of writing Chef programs or palindromes, sonnets or 4K demos? As a kind of physical education, these exercises might offer an extreme and precious template for ordinary life. But in fact they represent an opportunity to practice the deep, serious pursuit of circumscribed playgrounds. In so doing, the materials within a magic circle become defamiliarized, rendered from ground into figure. The syntax of C, or the quantity of ingredients in a recipe, or the selection of words used in a book can be seen directly rather than melting into the ether.

But in addition, the practice of working with constraints exercises the muscles of circumscription itself. The more frequently we create playgrounds and work within them, the more comfortable we will become in doing so when it matters. And when does it matter? Well, if you want to have fun, and if fun entails the discovery of something new in a familiar situation through the exercise of play, then there's not a moment when living with limits won't bear fruit. But more practical opportunities to transform constraints into culture arise from time to time as well.

Twitter is perhaps the only multibillion-dollar, publicly traded company ever to be built on the foundation of an Oulippian writing constraint. Users of the service can post text updates to their timelines, each of which can be no more than 140 characters in length. The social network was launched in 2006, and among its founders were Evan Williams, a creator of the blogging service Blogger (Twitter was sometimes referred to as "microblogging" in its early days), and Jack Dorsey, whose interest in passenger-car and emergency-vehicle dispatch routing had cultured an attraction to short-form status messages.

Smartphones didn't exist then in the same way that they do today—2006 was a year before the first iPhone, and the closest analogues were the BlackBerry and Palm Treo devices mostly used by executives and government officials for e-mail. Short message service (SMS) was the workhorse of mobile communication, and Dorsey thought there was potential for a short-form social network for "inconsequential information," meant to operate primarily over SMS.[34] The service took off after the 2007 South By Southwest conference, which has become something of a fashion week for new Silicon Valley products. By 2010, Twitter's users were sending sixty-five million tweets per day. The company went public in 2013, and for better or worse, its #hashtags and @handles and other conventions have become a staple of Internet-connected life for hundreds of millions of people who use it—and even more who encounter it in calls to action on television, in stores, and on print media.

SMS already had its own constraints, which Twitter built upon because it had to do so rather than because it chose to. SMS (we call them text messages nowadays) was conceived in the mid-1980s as a general mechanism for wireless carriers to send messages to and from handsets on the Global System for Mobile Communications (GSM) standard. GSM was built for voice telephony, not for text or data, and SMS had to fit within the limited bandwidth available at the time. GSM's Friedhelm Hillebrand originally proposed the 160-character limitation as a reasonable compromise between the network's capacity and the minimum possible space for sending a meaningful message. The first commercial SMS service launched in 1993, and by its twentieth birthday more than 8.5 trillion text messages were being sent and received annually.[35]

Dorsey and the Twitter team lopped 20 more characters off the 160 available on SMS to account for the sender's username. Over time, the 140-character limit inspired other conventions and constraints: the use of "RT" to indicate a "retweet" or forwarded tweet; "OH" for "overheard"; "ICYMI" for "in case you missed it." Twitter's material limits also popularized URL shortening services like bit.ly and shorturl, although Twitter has since rewritten URLs in tweets with their own service, t.co, which both shortens and records clicks centrally for use in data analytics and advertising sales. In fact, now that Twitter operates almost entirely separately from SMS, the average tweet sends a much larger data payload, including metadata about user location, the Twitter client who sends the tweet, media embedded in the tweet, other users mentioned, and so on.

More broadly, SMS has largely been replaced by other message services, from Apple's iMessage to WhatsApp to Facebook's Messenger, which don't impose length restrictions. Text messages used to be precious and costly—ten or twenty cents each when voice was the primary use for mobile phones—so you'd try to squeeze as much as possible into as few 160-character chunks as would carry it. In the United States, unlike most parts of the world, carriers often charge for received texts as well as sent ones, making unexpected conversations more than just a nuisance of superfluous beeps and buzzes. The relative difficulty of inputting a text message on a traditional twelve-key handset keypad also acted as a natural limiter to the length and frequency of the message, even after T9 predictive text made a whole generation of kids faster at texting than at typing.

Things are different today. Every smartphone has a full-size keyboard and data is the only rate carriers care about. It's metered, but only in aggregate, and the individual cost of messages has become infinitesimal and incalculable. And so we send as many messages as it takes to get our meaning across.

The same is true of Twitter. Precious, expensive short messages—from text messages to telegraphs—have become deemphasized, tweets included. If an idea doesn't fit in 140 characters, just spread it across more tweets at zero cost. New conventions for splitting long-form copy arose, among them the "tweet storm," a numbered list of individual components of a larger argument, Martin Luther's Ninety-Five Theses style. The deluge of messages from Twitter's half-billion or more users reduces the signal of each individual message. In an analysis of highly viewed viral tweets linking to news articles, *The Atlantic*'s Derek Thompson demonstrated that even with hundreds of thousands of impressions, only 1 percent of users who see a tweet click a link embedded inside to read the story to which it points.[36]

In other words, the constraint once central to Twitter's service has become increasingly incidental. As Twitter grew from an idea into a prototype into a company, the conceit of short inconsequential information gave way to the requirements of a publicly traded media company: growth of sign-ups, participation, and opportunities for ad sales and sponsorships to justify the valuations set by its investment hype. By early 2016, its flagging stock price worrying executives more than its loyalty to constraints, Twitter started to consider abandoning the 140-character constraint entirely.[37]

As a result, the experience of using Twitter feels far less like an exercise in constrained writing and more like every other social network—a business that fashions rationales to bring you back over and over, even when you might not want to return. Follower counts encourage the development of a privatized metric for worth and value. The prominent display of the numbers of likes and retweets, along with a whole panel of the app devoted to "mentions" produces the sensation that the MIT anthropologist Natasha Dow Schüll has called "the machine zone"—the numb hypnosis common among casino slot-machine addicts.[38] Today, the most likely SMS you'll receive from Twitter is one notifying you that someone, hopefully you, changed your account password.

But it's possible to imagine an even more constrained Twitter. Perhaps you could only send one tweet per day, for example. Or only follow 100 people at a time. Or if you never saw how many people follow or favorite or retweet you, or anyone else (this limitation in particular would reduce my own tendency to return to the service to check "how I'm doing" in the eyes of the twitterverse). Of course, none of these limitations correlate well with a contemporary technology company's business model, so you shouldn't expect to see them appear in an app update change list anytime soon.

Twitter didn't shift from Oulippian communicative playground to extractionist financial instrument. Rather, multiple overlapping playgrounds are at work in a system like Twitter—just as they are everywhere, all the time. By knowing which magic circles we are actively circumscribing, and which ones we are being circumscribed within, we can make better choices about the playgrounds we choose to create and occupy. Difficult though it may be in today's media ecosystem, it is possible to leave Twitter's playground entirely if its compulsive machine-zone features engulf its constrained writing and communicative features. Even so, all-or-nothing ultimatums are too ironic, anyway. Quitting Twitter or Facebook becomes its own identity, that of the conscientious objector who knows enough to cry Bartleby.

In an article indicting social networks' tendency to convert emotion and expression into corporate value, the critics Luke Stark and Kate Crawford connect *emoji*, the Japanese-import smileys and icons we now use in our texts and tweets, with big business. They remind us that Murray Spain, one of the entrepreneurs behind the 1970s smiley face, said of that icon, "This face is a symbol of capitalism." Emoji, Stark and Crawford conclude, "offer us a means of communicating that we didn't have before . . . , yet are also agents in turning emotions into economic value."[39]

It's true. But what isn't? Even seemingly unadulterated playgrounds like Chef and Oulipo are really only fantasies of purity. They too participate in other circumscriptions, reconfiguring themselves in collaboration with other entities—supermarkets and freight logistics, broadband and fiber infrastructure providers, snobbery and romantic Europhilia. The demand for a stamp of approval or denial is another symptom of ironoia, an entreaty for a fresh new blister pack that will enclose and present emoji or Twitter or whatever else as righteous and impervious to corruption or depraved and incapable of reform.

All the more reason to practice drawing and redrawing as many magic circles as possible, erecting as many playgrounds as we can muster. But as former figures become ground, we have the opportunity to reframe them, to put them to new uses by taking them seriously for their promise *and* their threat. Limits aren't limitations, not absolute ones. They're just the stuff out of which stuff is made.

SEVEN

The Opposite of Happiness

Mindfulness puts us at the center of our fates.
But mistaking the world for our world amplifies the anguish
of ironoia. Instead we must learn to respect the things
around us for what they are rather than for what they lack.

E VERY MARCH, MORE THAN TWENTY-FIVE THOUSAND GAME makers gather in San Francisco for their annual industry conference. Like endodontics or insurance underwriting, games are a profession, and the Game Developers Conference offers a forum for its creators to share and acquire knowledge about their practice.

At the 2010 meeting, a schism erupted between "traditional" game developers, who make the sorts of console and casual games we've come to associate with the name "video games," and so-called "social game" developers, who make games for Facebook and other social networks. It was a storm that had been brewing for a few years, but the massive success of Zynga's FarmVille along with the company's malicious and braggart attitude about that success had made even the most apathetic of game developers suddenly keen to defend their craft as art.

FarmVille seems harmless enough on the surface: you build a farm, plant crops, raise livestock, and then harvest them at specified intervals for virtual profit, which you can reinvest in continued development and expansion of the farm. But Zynga's affronts were real: lead-generation advertising scams encouraged players to buy or subscribe to questionable services in exchange for FarmVille currency; viral loop customer acquisition predicated on spamming your friends and acquaintances for in-game incentives; and overtly compulsive gameplay, built around scheduled "withering" of crops that forced players to organize their schedules around predetermined play sessions set at intervals just long enough to forget about but not long enough to make them fit into ordinary schedules. It was ironic, I suppose, for a group of professionals often criticized for mainly facilitating the power fantasies of adolescent boys suddenly to claim the moral high ground against a start-up offering extortion-rich, simulated cartoon farming for soccer moms. Nothing brings a group of bandits together like a new bandit in town.

In July of that year, my colleague Jesper Juul invited me to take part in a game design theory seminar he was running at New York University, which he provocatively titled "Social Games on Trial." The Finnish researcher and social game developer Aki Järvinen would defend social games, and I was to speak against them.

As I prepared for the NYU seminar, I realized that theory alone might not help clarify social games—for me or for anyone in attendance. Instead of pontificating from a bully pulpit as if I were somehow above it all, I decided to make an example that would act as its own theory. In the case of social games, I reasoned that enacting the principles of my concerns might help me clarify them and, furthermore, to question them. So I decided to make a game that would attempt to distill the social game genre down to its essence.

The result was Cow Clicker, a Facebook game about Facebook games that completely consumed my life for the ensuing year and a half. The game itself was simple enough. There was a cute picture of a cow, which players could click every six hours. Each time they did, they would receive one point, called a click. As was common in Facebook games of the time, players could invite friends to join in by adding them to a "pasture." When any of one's pasture-mates clicked, all players in the pasture would receive a click too. Players could purchase in-game currency, called "mooney," which they could use to buy more cows or to skip the click timer. There was more—much more, embarrassingly more—but that's enough to get us started.

Whether I like it or not, Cow Clicker was easily the most successful game I have ever produced. Many tens of thousands of people played it, but many more played with the idea of it. In most circles, I am now most easily introduced as "the Cow Clicker guy." If some contemporary Homer were to mnemonize me, surely some cow pun would become my epithet.

As 2010 gave way to 2011 and beyond, I was hoist on my own petard, as compulsively obsessed with running my stupid game as were the players who were playing it. I added cow gifting; an iPhone app ("Cow Clicker Moobile"); a children's game ("My First Cow Clicker"); a "cowclickification" API that both mocked and embraced the then-rising trend of gamification; "cow clicktivism," a partnership with my radical Italian game-designer friend Paolo Pedercini, which allowed players to buy a package deal: a grievous, emaciated clickable cow and a real cow sent to the third world via Oxfam Unwrapped.

When the time finally came time to end it, Cow Clicker having bested me, I launched a bovine alternate-reality game (a kind of paranoia-fiction treasure hunt) played on four continents that revealed the coming "cowpocalypse." Once the last clue was retrieved from the back of a Sydney street sign or the bottom of a London pub table or the poster board of a Brooklyn vegan bakery, the pieces assembled, Voltron-style, and a huge clock appeared. Nobody knew what it meant, but it was the countdown to Cow Clicker's rapture. Armoogeddon.

Writing about the game in *Wired* magazine the following year, Jason Tanz told the story of Jamie Clark, a military spouse living on Ellsworth Air Force Base, in South Dakota. She had close friendships with her fellow clickers. "I don't meet a lot of people who discuss politics and religion and philosophy, but these people do, and I like talking to them. I'd rather talk to my Cow Clicker friends than to people I went to school with for twelve years." Tanz draws a tough conclusion, one I can't help but accept: "It's a common refrain among dedicated Cow Clickers, who have turned what was intended to be a vapid experience into a source of camaraderie and creativity." He continues:

> It may be that Cow Clicker demonstrates the opposite of what it set out to prove and that social games, no matter how cynically designed, can still provide meaningful experiences. That's how Zynga's Brian Reynolds sees it. "Ian made Cow Clicker and discovered, perhaps to his dismay, that people liked it," Reynolds says. "Who are we to tell people what to like?" Gabe Zichermann, a gamification expert, also dismisses Bogost's critique of Zynga's games. "Other gamers may think FarmVille is shallow, but the average player is happy to play it," he says. "*Two and a Half Men* is the most popular show on television. Very few people would argue that it's as good as *Mad Men*, but do the people watching *Two and a Half Men* sit around saying, oh, woe is me? At some point, you're just an elitist fuck."[1]

I had responded to this idea almost a year earlier, in a "rant" session at the 2011 Game Developers Conference. My contribution was entitled "Shit Crayons."[2] In it, I compared Cow Clicker players to the Nobel Prize–winning Nigerian poet Wole Soyinka. While imprisoned after the coup that led to the Nigerian Civil War of 1967, he composed poetry and plays from his cell using whatever writing material he could find. How resilient is the human spirit that it withstands so much? No matter the trials to which we subject people, nevertheless they endure. They thrive even, spinning shit into gold.

The cowpocalypse finally arrived on the evening of September 7, 2011. Frantically working at a makeshift desk in my den, I flipped a few virtual switches I had set up weeks earlier, and all the cows disappeared. In their place was empty grass. Tanz explains better than I could:

> They have been raptured—replaced with an image of an empty patch of grass. Players can still click on the grass, still generate points for doing so, but there are no new cows to buy, no mooing to celebrate their action. In some sense, this is the truest version of Cow Clicker—the pure, cold game mechanic without any ornamentation. Bogost says that he expects most people will "see this as an invitation to end their relationship with Cow Clicker."
>
> But months after the rapture, Adam Scriven, an enthusiastic player from British Columbia, hasn't accepted that invitation. He is still clicking the space where his cow used to be. After the cowpocalypse, Bogost added one more bedeviling feature—a diamond cowbell, which could be earned by reaching one million clicks. It was intended as a joke; it would probably take ten years of steady clicking to garner that many points. But Scriven says he might go for it. "It is very interesting, clicking nothing," Scriven says. "But then, we were clicking nothing the whole time. It just looked like we were clicking cows."[3]

I'm sure you'll recognize the ironic maneuvers in Cow Clicker. Layers and layers of reference and repackaging. Game turned game about game, turned host for commentary upon its own success, turned sacrificial lamb through simulated martyrdom. So urgent was it to me that the game perform the critique I'd set out for it that I was unable to admit that its features far exceeded mere censure or success. Reality turned out to be much weirder. Despite my intentions, Cow Clicker was just as much of a playground as anything.

Others continued working in that playground even after I abandoned it. In 2013, Julien "Orteil" Thiennot created Cookie Clicker, a browser game in which players click on a cookie to bake cookies. Eventually, those cookies can be used to buy resources that bake more cookies—grandmas, farms, factories, and so on. Once the player has mustered these automated bakers into service, clicking becomes superfluous. The game bakes its own cookies, and far faster than you could ever click them into existence.

Orteil had discovered the logical conclusion of Cow Clicker, the place I didn't think to take it: a game you don't even have to play. People rejoiced! Within a month, more than 200,000 people were "playing" Cookie Clicker every day.

It didn't stop there. For whatever reason, the conditions were right for not only one or two clicker-games, but a whole genre of them. Cookie Clicker isn't a game for a human, but one for a computer to play while a human watches (or doesn't). What better way to relieve the anxiety and compulsion of social games and of apps and websites and digital life more generally? Eventually the genre acquired a name, "idle games," although to my glee some folks still call them "cow clickers." Now there are dozens of them, not only gags but earnest titles with budgets and players making real money for real creators, which they're using to pay real mortgages and to buy real Taffyta Muttonfudge action figures.

Sounds stupid, right? But so does Cow Clicker or FarmVille or Twitter or sonnets or anything, really, when you defamiliarize it. Stupid as in stunned, arrested, trapped in a *Matrix* freeze-frame between ground and figure, ready to be seen more closely. Foolish—and fun. The purpose of play is to make things seem stupid, so you can determine what they do and what you can do with them. And with stupor comes amazement, terror even. There is terror in fun, because it shows us what we fail to see, what we refuse to see, what we fear to see.

FUN IS THE OPPOSITE OF HAPPINESS

It's not that I was wrong about Cow Clicker, not exactly. It did what I thought and hoped it would do, but it also did more than that. The glory of Cow Clicker isn't that it succeeded or failed as a critique of Facebook or FarmVille, but that it was so much bigger than just that critique, whether I liked it or not—whether I even noticed it or not. The universe doesn't care how gratified or offended we are at the reception of our creative output—nor at how much gratification we derive from a Walmart.

It's a far cry from Marie Kondo's idea that we ought to surround ourselves with objects that spark joy. But it's no surprise that joy would sell more books (and destroy more apparently joyless socks, skirts, books, and other trifles) than would more ambiguous sensations. For two centuries now, optimism and happiness have triumphed in replacing for the lost the guidance of spiritualism.

Positive psychology started much earlier than Csikszentmihalyi's flow. Optimism got its modern start in Transcendentalism of the nineteenth century, the movement of Ralph Waldo Emerson and Henry David Thoreau. Transcendentalists celebrated individualism, naturalism, and reason over the corrupting forces of society, especially religion and politics. As Emerson put it in his 1836 essay "Nature," generally considered the movement's founding text, "As fast as you conform your life to the pure idea in your mind, that will unfold its great proportions."[4]

A less poetic and more marketable version of self-reliance emerged in the mid-twentieth century, as prosperity, industry, and unambiguous military might made the human will seem indefatigable. Norman Vincent Peale's 1952 book *The Power of Positive Thinking* became the urtext of the self-help genre. Optimism became the self-fulfilling prophesy it is today, an unverifiable but effective trump card against all obstacles. Those who succeed appeal to a positive outlook as a component of their success, while those who fail are deemed to have been insufficiently optimistic, no matter how much positive thinking they may have thought themselves to be embracing. Optimism creates a paradox similar to Barry Schwartz's surplus. As Barbara Ehrenreich puts it in her book *Smile or Die: How Positive Thinking Fooled America and the World*, "If optimism is the key to material success, and if you can achieve an optimistic outlook through the discipline of positive thinking, then there is no excuse for failure."[5]

In Peale's day, optimism and happiness were anecdotal matters, which contributed to their exploitation as self-help snake oil in the ensuing decades. But today, happiness has become a statistic that some psychologists and neuroscientists believe they can measure. A ghastly infographic created as an advertisement by the startup Happify celebrates the simplicity of the matter.[6] "Happiness," the company's shill reads, "is a combination of how satisfied you are with your life + how good you feel on a day to day basis." Readers who scroll down Happify's webpage will delight in knowing that 40 percent of happiness is controlled by your thoughts and actions, 10 percent by your circumstances, and 50 percent by genetics, that favorite trump card of contemporary scientism.

Those widely cited stats come from Sonja Lyubomirsky, one of many researchers who aim to replace the anecdotal claims of Peale and his heirs with clinical and scientific explanations. And yet in her popular book *The How of Happiness*, Lyubomirsky reveals that her understanding of happiness is no less anecdotal and arbitrary, never mind the surveys and interviews. "I use the term *happiness*," Lyubomirsky writes, "to refer to the experience of joy, contentment, or positive well-being." Then the other shoe drops. "Academic researchers prefer the term *subjective well-being* (or simply *well-being*) because it sounds scientific and does not carry the weight of centuries of historical, literary, and scientific subtexts."[7] You'll be forgiven if you think this supposedly scientific-*sounding* term might also be construed as a clever rhetorical flourish meant to be draped luridly over its seductive subject.

In 2013, an otherwise unknown University of East London master's student named Nick Brown helped debunk the dubious mathematics of another popular and supposedly scientific claim about the mechanics of happiness. Barbara Fredrickson and Marcial Losada had published an article in the respected journal *American Psychologist* called "Positive Affect and the Complex Dynamics of Human Flourishing," which proposed that a simple ratio of positive to negative emotions (2.9013 to 1) separated people who "flourish" from those who "languish." Brown teamed up with the famous mathematician and infamous academic prankster Alan Sokal, and the two wrote a response, "The Complex Dynamics of Wishful Thinking," which argued that the mathematics Fredrickson and Losada had deployed amounted to an elaborate mathematical parlor trick with no basis other than an abstract appeal to mathematical truth.[8]

It's not only positive psychology that's guilty of contorting happiness to serve its own ends. Snake oil and happiness have always been intertwined. In the first episode of *Mad Men*, Sterling Cooper creative director Don Draper offers this explanation of the ad business:

Advertising is based on one thing: happiness. And do you know what happiness is? Happiness is the smell of a new car; it's freedom from fear; it's a billboard on the side of the road that screams with reassurance that whatever you are doing is okay. You are okay.

Though the TV show is fictional, Draper's summary is accurate: fear is a prime mover in marketing. The canonical example is Listerine's 1920s ads for mouthwash, which transformed bad breath into *halitosis*—a ghastly sounding condition sure to exile even the most comely of man or woman. Luckily, Listerine offered a cure for the disease it had invented. The product's sales increased sevenfold over the next decade.[9]

I suppose there's nothing wrong with believing in positive psychology, optimism, happiness theory, the science of well-being, or whatever else you wish to call it—after all, millennia of human beings have fashioned explanations for contentment, and there's no reason to shame people for using techniques they find helpful. But explanations become squirrelier when scientific methods are used as a justification for anecdotal appeals to vanity. And the scientific method uses the very same appeals that self-help gurus and snake-oil salesmen have used to sell positivity, like Listerine sold a cure for halitosis.

Almost every supposedly scientific case for happiness would be more compelling as a philosophical one. Fredrickson and the rest make perfectly agreeable rhetorical arguments, as Lyubomirsky's struggle to choose and explain her language attests. But in the secular age, science has replaced religion as sanction. Jackson Lears explains the consequences of this scientism in *The Nation*:

It is not, to be sure, an outlook based on the scientific method— the patient weighing of experimental results, the reframing of questions in response to contrary evidence, the willingness to live with epistemological uncertainty. Quite the contrary: scientism is a revival of the nineteenth-century positivist faith that a reified "science" has discovered (or is about to discover) all the important truths about human life. Precise measurement and rigorous calculation, in this view, are the basis for finally settling enduring metaphysical and moral controversies.[10]

By treating self-reported survey data and its derived models, statistics, and ratios as indelible truth, positive psychology unwittingly stifles the playful ambiguity that comes from metaphysical and rhetorical argument. At best, we're back to Mary Poppins's logic: happiness makes you happy, so put some happy in your happiness so you can be happier. At worst, we've suffocated all the play in human experience and replaced it with a series of metrics and measures, ready to be formalized into future Google smartwhatevers or Apple cortex apps, whose sensors and feedback systems and machine learning algorithms will optimize our future experiences to match the "proven" positive psychological truth du jour.

Happiness science or positive psychology mostly get us back to the same place we started, and the same place we've been with happiness since utilitarianism: appeals to our own self-interested outcomes. It returns the spoils of every hypothetical interaction to a self-reported measure of internal delight, whether you call that delight happiness or joy or contentment or flourishing or well-being.

Even Marie Kondo's animist uncluttering manifesto is a breath of fresh air compared to positive psychology. At least she doesn't make appeals to some transcendental power in order to justify why you should lay your socks flat or donate your camping gear to the thrift store. But still, the spirit of her instructions is essentially selfish, too. How, KonMari asks, can I create the environment most suited *to me*? What can this particular object do *for me*? Only once a thing is spared the incinerator does it become a candidate for the animist respect that invites socks to lounge flat so that they might rest. Move your shit so I can put my stuff down.

The philosopher Ada S. Jaarsma compares Don Draper's version of advertising's "freedom from fear" to the existentialist symptoms of bad faith—stagnated, seeking assurance in being at rest rather than in motion toward the future.[11] Eventually, near the end of her little bible on tidying, Kondo adopts the same position: "I can think of no greater happiness in life than to be surrounded only by the things I love."[12]

■ ■ ■

FUN IS THE opposite of happiness. It seems unintuitive, until you look at the matter from the perspective of happiness rather than play. In his book *The Antidote: Happiness for People Who Can't Stand Positive Thinking*, the journalist Oliver Burkeman explains a well-documented principle of psychological sorrow. "Hedonic adaptation" is the tendency to grow acclimated to something pleasurable, such that it ceases to deliver the joy it once did. "Whether it's as minor as a new piece of electronic gadgetry or as major as a marriage," says Burkeman, it "swiftly gets relegated to the backdrop of our lives."[13]

Burkeman connects hedonic adaptation to the third-century BC philosophy of Stoicism. The Stoics advocated for establishing agreement with nature, a part of which entails the rejection of undue optimism. While he never calls it by this name, Barry Schwartz's endorsement of satisficing is a kind of stoicism in which one reduces hopes for the outcome of choices by reducing the options for choice in the first place. Burkeman refers to a famous claim of the Stoic Epictetus, whose philosophical position rests on rationality. Here's the classic example from Epictetus's *Discourses*: "At the very moment you are taking joy in something, present yourself with the opposite impressions. What harm is it, when just as you are kissing your little child, to say: 'Tomorrow you will die,' or to your friend similarly, 'Tomorrow one of us will go away, and we shall not see one another any more'?"[14]

It's no accident that Burkeman uses the term *backdrop* to characterize hedonic adaptation. The reduced pleasure we derive from something arises when that thing becomes ground, where things go unnoticed even though our experiences rely upon them. Only by making them figure can we even see them again such that we might start appreciating them anew. This kind of appreciation is what Epictetus means by his startling suggestion that parents imagine the imminent death of a child in order to counteract the narcotic effects of joy. He's suggesting that you fashion a quick, if morbid, playground around your parental affection in order to appreciate it anew.

We can reframe the Stoic defamiliarization in terms of the machinery of play and its exhaust in the form of fun. Colloquially, we elide fun and joy or happiness or pleasure, so that it would make no sense to call pondering the untimely death of a child "fun." But if fun instead refers to the discovery or rediscovery of a feature of a thing outside of you, then it's no overstatement to cite Epictetus's example as an exercise in fun. Its purpose is not to produce joy, but the opposite: the recognition of the unrealized potential of another. A playground in which to rehearse love by pondering its hypothetical loss.

Anyone who has children or pets or even socks that spark Kon-Maric joy has probably experienced a sensation of this kind. Watching a baby sleep or coursing one's hand over the upholstery of a favorite chair unleashes a flood of affect that overwhelms reason, welling up involuntarily in your throat and eyes. No one would call this feeling *joy*, but sorrow isn't the right word for it either. It's the opposite of ironoia, an incontrovertible respect for the thing, a recognition that you are small in its presence. Fun doesn't produce joy as its emotional output, but tenderness instead. Affection and warmth and sympathy. It marks the encounter with another being—human, animal, vegetable, plastic, Optimus Prime piñata—that asks and expects nothing back, but only seeks to understand something more about what one faces. Happiness and joy are selfish goals, but play and fun are selfless ones.

Even KonMari contains the seeds for a less selfish and more generous version of joy. Just before instructing her tidiers to ask whether candidates for disposal spark joy, she says something else, a precursor to the question: *"Look more closely at what is there."*[15] The emphasis is Kondo's, so she clearly thinks it's an important point. But from the perspective of fun instead of joy, the purpose of the order is misplaced. Kondo wants closer attention to lead to a more deliberate measure of one's affective attachment to the thing in question, leading to a verdict on its worthiness. Something insidious is going on here, and it's easy to miss it, thanks to the cute, chipper matter-of-factness of Kondo's writing. A Hello Kitty executioner.

NOT MINDFULNESS BUT WORLDFULNESS

But there's another interpretation of Kondo's invitation to commune with objects, and one from the same tradition of Eastern animism that motivates her invitation to respect your socks. The Japanese-American designer and computationalist John Maeda, formerly a professor at MIT and the president of the Rhode Island School of Design, discusses the Shinto tradition of seeing a living spirit in everything. *Aichaku* is a Japanese term that is usually translated as *attachment*. But, as Maeda explains, its two kanji characters *ai* (愛) and *chaku* (着) mean "love" and "fit," respectively. "Love-fit," Maeda suggests, is a "symbiotic love for an object that deserves affection not for what it does, but for what it *is*."[16]

Sometimes this respect for objects in Eastern spiritualism still returns its spoils to the self. Buddhism, for example, holds that attachment (*upādāna* in Sanskrit, "fuel" rather than "love-fit") is the primary cause of worldly suffering. Buddhism holds that all things are ultimately impermanent, and attachment is the self-deception of permanence. To combat attachment, Buddhism exhorts its adherents to resist developing certainties about how things are, should, or shouldn't be. Instead, the Buddhist seeks a more general peace with the universe. The mantras and meditative practices associated with Buddhism offer mechanisms for developing and maintaining the peace of detachment. Like physical exercise, it must be conducted regularly.

The physical therapy that play and fun provide as a salve for ironoia are not spiritual practices akin to Buddhism or Shinto. If anything, fun is a far messier and less minimalist approach to living with things. Instead of removing unnecessary or joyless things from your presence, play invites you to consider your surroundings as a vast domain of essentially limitless meaning and potential. Play and fun elicit a forthrightness and even an aggression for which Buddhism has no patience. Playgrounds are places where we dig deep, where we mess things up and tear them asunder—ourselves included—in order to discover what else is possible. Not for useful ends, necessarily, nor necessarily for useless ones either.

The upside of irony is that it starts with the right sentiment: "anything whatsoever has potential." The happiness advocate adds one codicil, ". . . and that potential must return joy to the human who interfaces with it." And then, the ironoiac adds another, ". . . but it will always prove insufficient." Happiness is in cahoots with ironoia: inwardly focused, deployed as a snare to extract and return worldly resources for purification into joy or glee or satisfaction, and thus a source of the anxiety that arises once that outside world proves to be less resistant to refinement than fossil fuels or sugars or carbohydrates. The comfort that happiness would supposedly provide proves both harder to secure and less likely to do what it promises even once obtained.

To challenge the happiness rhetoric's failure to produce the comfort it claims to proffer, Burkeman turns to Alan Watts, the British philosopher largely responsible for bringing Buddhism, Taoism, Zen, and other staples of Eastern thought to the West in the mid-twentieth century. In 1951, the year before Peale's book on positive thinking hit the shelves, Watts published *The Wisdom of Insecurity*. He argues that the twentieth century marked the divergence of two paths for seeking comfort, religion, and science. As far as the options go, not much has changed in more than half a century since, even if science's stock has risen markedly, to the point that extreme scientism looks more and more like religion.

Watts offers a third option: rather than choosing between Heaven and Earth, Nature and Spirit, we can focus on impermanence as a compass bearing. Happiness, positive thinking, religion, and science all seek fixed answers. Whether in the form of rites, virtues, and preparations necessary for admission into the Kingdom of Heaven, or via a deduction like Fredrickson's ratio of flourishing, contentment is posed as a question that can be answered. A certainty. For Watts, the pursuit of certainty entails a hidden maneuver: the attempt to separate ourselves from the rest of the world, and in so doing to prevail over all that other mere matter, whether via faith or reason.

Even in philosophy humans have sat at the center of being, ethics, and aesthetics since the Enlightenment, when Immanuel Kant imagined a Copernican Revolution in philosophy to mirror the one in cosmology. It was from this common origin in Enlightenment rationalism that human culture spread in the different directions Watts judged insufficient. Science broke down the biological, physical, and cosmological world into smaller and smaller bits in order to understand its truth, which would then be deployed to service the betterment of humankind. Religion redoubled its investment in God via a new dedication to reason, both through hybrids like God-as-watchmaker deism, and via the religious revivalism that rose up against the new era of reason. In both cases, the reign of mankind on Earth remained absolute, its terrestrial matter at our absolute disposal thanks to the pending eternity we would find in heaven. For its part, philosophy concluded that reason could not even explain external objects but only experience itself—this was Kant's position, one that still hasn't outworn its influence in epistemology and metaphysics. And it was the position in more populist cultural debates, too, where all reality gets strained through the sieve of culture, religion, science, politics. Everything whatsoever becomes but the expression of human will or ideology.

The sociologist Bruno Latour wraps all of it—religion, science, philosophy, and more—into one historical era: modernity. Starting in the Renaissance, human cultures began dividing and "purifying" the domains of nature and culture from one another.[17] But this modern idea is really a misconception; instead, things are *hybrids* of nature and culture, mankind and world, society and individualism. We are all bound up with so many other actors, as Latour calls them: roadways and chlorofluorocarbons, mature woods and IKEA boxes, global banking networks and the wayward birds that clog jet turbines.

Worse, the modern misconception is a positive feedback loop, a Chinese finger trap in which we only become more imbricated the more we try to escape it. Climate change is a hybrid created by the fusion of human progress with planetary fragility, a collaboration underwritten by myriad political, industrial, and economic circumstances. But when we insist on the inevitable defeat of nature by science, we end up with even more hybrids, and even more hybridized ones: robotic car services that promise to end the reign of the passenger automobile, or the dream of uploading ourselves to powerful computers that will simulate human consciousness. Latour advocates for a "parliament of things" that accepts hybridization and ends the false dream of purification.[18] Among those who have carried Latour's torch is the philosopher Graham Harman, whose revision of the Heideggerian idea of readiness-to-hand and presence-at-hand applies to the relationships between objects of all sorts we visited earlier. It's not only that people find hammers to be ready-to-hand when in use and present-at-hand when broken; it's also that hammers and nails, hammers and hooks, hammers and rust have the same relationship.

That concept leads us back to Alan Watts. When we feel insecurity, we are really feeling the wish for our own permanence. "We do not actually understand that there is no security," writes Watts, "until we realize that this 'I' does not exist."[19] It's a decidedly Buddhist statement whose new age bell ringing might make your eyes roll, but Watts is really saying the same thing as Latour and Harman: we humans are not separate from the world, but a part of it, just another thing among all the other things, from ketchups to ketch boats, RAV4s to ravines.

And as Burkeman also divines, mistaking the world for *our* world is another name for ironoia. Thinking of ourselves as centered bodies drawing meaning and contentment toward ourselves like gravity to planetary bodies, bodies that *deserve* something from everything. But our refusal to relinquish faith in the "I" that does not exist, says Burkeman,

> . . . explains in the most complete sense why our efforts to find happiness are so frequently sabotaged by 'ironic' effects, delivering the exact opposite of what we set out to gain. All positive thinking, all goal-setting and visualizing and looking on the bright side, all trying to make things *go our way*, as opposed to some other way, is rooted in an assumption about the separateness of "us" and those "things." But on closer inspection this assumption collapses.[20]

Embracing the play made possible by limitations turns happiness inside out. When you circumscribe a magic circle around something, you inaugurate a different attitude from the modernist, from the self-help happiness seeker, from the ironoiac. Instead of asking for an affective return on investment transferrable to specific psychological outcomes, you submit yourself to the same system as the elements you have drawn into your magic circle. You become a thing like them, imbricated, bound up, twisting, dancing, trying to make the Rube Goldberg machine that is you-and-them work, turning the crank not because it does something useful, but just to see what it does.

■■■

PLAY INVITES US to draw an overdue conclusion: that the potential meaning and value of things—anything: relationships, the natural world, packaged goods—is *in them* rather than in us. Play is not a kind of self-expression, nor a pursuit of freedom. It is a kind of creation, a kind of craftsmanship, even. By adopting, inventing, constructing, and reconfiguring the material and conceptual limits around us, we can fashion novelty from anything at all. Although they refer to *poiesis*—the making that grounds poetry—instead of play, the philosophers Bert Dreyfus and Sean Kelly come to a similar conclusion about finding meaning in a secular age: "The task of the craftsman is not to *generate* the meaning, but rather to *cultivate* in himself the skill for *discerning* the meanings that are *already there*."

In his rejoinder to the cult of happiness, Oliver Burkeman introduces us to Paul Pearsall, a neuropsychologist who contends that most people live "awe-deficient lives." To describe the immersion in experience, Pearsall suggests the neologism "openture"—the opposite of closure. Openture is the immersion in experience rather than its rejection. As Burkeman describes it, openture "is not certitude or even calm or comfort, . . . but rather the 'strange, excited comfort of being presented with, grappling with, the tremendous mysteries life offers.'"[21]

Isn't openture just another name for play? There is a humility in limitation, for it requires us to treat things *as they are* rather than *as we wish them to be*. Such a shift is required if we hope to escape the tyranny of restraint in order to construct a life in which we take the world around us more seriously by paying greater attention to its contents.

Instead of seeking greater happiness within ourselves, we should pursue a greater respect for the things, people, and situations around us, learning to take them for what they are rather than for what they lack. The best way to combat the anxiety of a world whose contents might disappoint is to decenter ourselves from that world's concern. And despite the inspiration we might find in Buddhism and related Eastern practices, such ancient catalysts are not enough to help us reconcile letting go of ourselves with the enticements of the contemporary world. Ours is a world where BMWs and Bagel-fuls, FarmVille and Nyan Cat confuse the apparent simplicity of letting go of attachment. The solution to our ongoing boredom, anxiety, and ironoia cannot be found in us, through increasing investments in mindfulness, but anywhere and everywhere else.

We need a new commitment to *worldfulness*, an agreement to remain open to the infinity of things that are *not* our own feeble minds trying so hard to shape the world to our hopes, fears, and assumptions. The opposite of happiness isn't sorrow, or misery. It's esteem. It's reverence. Nothing is here for you, not you alone, even if you put it there. And thus, around every turn in the woods, sidled up at every barstool, stacked on every supermarket endcap, we find potential collaborators, each ready to show some part of their capacity when operated, or observed in operation, or simply pondered from afar.

Living with Things

I N 1954, UNITED STATES AIR FORCE LIEUTENANT COLONEL John Paul Stapp strapped into the exposed cockpit of a rocket sled named Sonic Wind and launched himself down a two-thousand-foot railroad track improvised into the empty dust of Holloman Air Force Base, near Alamogordo, New Mexico. He reached a ground speed of 632 miles per hour before meeting a trough of hydraulic brakes, which decelerated the sled to a halt in one and a half seconds. In so doing, Stapp withstood over forty times the force of gravity.

Colonel Stapp's work on the effects of deceleration had a profound influence on both military and civilian machinery. In addition to inspiring standards for fighter jets, his work on the common shoulder strap and lap belt was both theoretical and applied. He pioneered the first use of crash-test dummies and advocated heavily for automobile seatbelts.

I thought of Stapp when the nylon webbing of my car's shoulder strap stopped short as I piloted my car off Howell Mill Road and into the Walmart parking lot that would become the site of my big-box excavation. *Ground* names the underlying layer of stuff we rely upon without noticing, as ready-to-hand tools, as infrastructure. Not only to ports and freeways and roads and electrical grids and sewers, but also to seatbelts and Walmarts. In which case, Stapp helped underwrite our ability to stock up on Scoop Away kitty litter and Sea Best canned pasteurized crabmeat no less than Dwight Eisenhower or Thomas MacDonald. If Robert Moses was the architect of suburbia, John Paul Stapp was its day laborer, the guy nobody thinks about but without whom all the freeways and avenues and mini-malls would be impossible.

The same summer that Colonel Stapp withstood the highest known measure of force voluntarily experienced by a human, an eighteen-year-old man drove home alone through the dark, quiet streets of Milwaukee, past the scrap recyclers and metal shops of North Division. It was very late at night, and the young man was returning from a date. Because he was young, and because he was tired, and because no radio broadcast could provide a jolt of one of that year's pop standards by the Four Aces or the Crew-Cuts, he fell asleep at the wheel. A deceleration less well planned than Sonic Wind's inadvertently delivered his automobile to the neighborhood's scrapyards.

Fourteen more years would pass before Stapp's seatbelts would become a required feature in cars sold in the United States. My father would recover from most of the injuries he'd sustained after his unharnessed frame was thrown through the windshield and onto the pavement. But when he finally woke up a month later, one trauma would remain forever. It wasn't the crash itself that would permanently damage his vision, but the brain surgery that probably contributed to saving the rest of him, and thus indirectly underwriting me.

It's hard to know precisely what happened—perhaps a nick of the scalpel against the optic chiasm, perhaps an unknown consequence of a then-experimental procedure, or perhaps something inexplicable, something spectral. Surely the story has been mythologized anyway in its retelling, the way that tragedy always betrays drama and theatrics. No matter. The result would be a lifetime of near blindness, uncorrectable with optical prosthetics due to its source, withdrawn deep into the cranium, invisible.

I have breathed the hot exhaust of my dad's disability my entire life, like secondhand smoke. There were always some things he'd admit he couldn't do (like drive a car), but his dogged independence and overall bullheadedness contributed to his stalwart refusal to refuse to have a go at things.

It's a virtue as much as a vice. And probably the same character that would insist that a tie was brown instead of green, or that a public toilet had been marked men instead of women, or that he didn't know I was there when he ran into me—probably those same traits also helped him persist in college, eventually earning a doctorate in psychology the same year the seatbelt became mandated, a time long before the Rehabilitation Act and the Americans with Disabilities Act made special accommodations for students like him mandatory. In a charming paradox, he practiced as a clinical vocational rehabilitation psychologist for three decades. Unable to see outward adeptly, he looked inside those he couldn't see.

Because nothing was facile, no device sufficiently present to qualify as equipment, everything is topped with a frothy head of mystery:

> When traversing a neighborhood block or parking lot, a constant vigilance for curbs, stoops, stumps, and other imperfections, which my mother or I might call out and which my father would selectively ignore with ironic assurances, "Yes, I see it, I see it."

As if it were a game rather than a tool, the playful counting of steps and stairs, both aloud and with impressively overstated footfalls, which also serve to combat the constant influx of rogue asphalts, cements, twigs, and other detritus that would go unnoticed.

A ubiquitous pocket magnifier to investigate details of possible interest. But unlike the large lens of a Sherlock Holmes, my father's investigations yielded more commonplace conclusions: the sale price of Coca-Cola six-packs, or the current position on an FM dial lost in static, or the location of a rogue Excedrin tablet on a white countertop.

When at a restaurant, my mother reading a litany of menu items aloud—far too loud, I might add—including every detail of its preparation: grilled sea bass, roasted potatoes, forest mushrooms, spinach, orange beurre blanc; grilled salmon, whole grain mustard cream sauce, fried polenta, green beans, sautéed arugula, roasted peppers; heritage ranch grilled beef filet, Cambozola cheese, mushroom risotto, asparagus, Nebbiolo wine sauce.

As a kid, Mom's oration was a nuisance. Even as an adult, it's a nuisance. Habitual eccentricity and obstinacy crafted a unique being in my parents. One that works according to a logic hidden within their molten core. As their son, I was particularly accustomed to interpreting the lava flows that seeped therefrom, often offering direction for the worldly souls who happened upon them. I'd always considered this withdrawal a liability, a defect—a chronic isolation, a refusal or inability to normalize behavior to the standards of ordinary society, impairment or no.

But in that irritation, something was hidden in plain sight. My daily life as a child implemented the kind of physical education Pasolini had in mind. Without knowing it, I had grown up assimilated with mundane objects. Every day was a day full of innumerable playgrounds, thrust upon me whether I was willing to manipulate their contents or not. I was uncompromised by the conventions of ordinary

use—or even ordinary sense. A ballad of social perplexity evolved into
a tribute to the surprising, understated infinity of existence:

> Tree roots that evolve mountains out of concrete
> The unpainted edges of uneven stairs
> The brown buttons on hotel ice machines which blend into
>   the tan bezel that contains them
> The edge of the water-spotted flatware that precariously tips
>   the coffee cup
> The dots of halftoned ink that smear to fashion the letterforms
>   on newsprint
> The polyp that distinguishes a tangelo from a navel orange
> The paralyzing black of a darker than usual darkness

This was my physical education; these were the material phe-
nomena of my social upbringing. More cheeseburger-flavored Pringles than Super 8 camera.

There is something profoundly humble about things in the wild.
It's tempting to mistake that humility for tawdriness or for avarice,
to think of ordinary stuff as the detritus we must sift through or reject on our quest for beauty or justice. But what if we make just the
opposite assumption? What if we believe, even through pretense if
need be, that everything has the capacity to educate us in the flesh
and spirit, as Pasolini suggests? Not only clarinets and cameras, but
also adult diapers and pour, store, and pump jugs? In that case, carefully choosing the objects with which we surround ourselves as a
way of optimizing meaning or virtue is futile.

■ ■ ■

I CONFESS THAT I had forgotten about my childhood among
things, the one underwritten by my father's constant attention

to and precariousness among them. I returned to it only by accident. Some years ago, I learned how to program the Atari Video Computer System, the 1977 console that made home video games popular. My colleague Nick Montfort, the neo-Oulippian you met in chapter six, and I were writing a book about the relationship between the hardware design of the Atari and the creative practices its programmers invented in those early days of the video game.[1] The project demanded that we really understand how the machine operated, not only how its games looked and played. It's the opposite of ironizing the Atari as a vintage, nostalgia object: grasping it as a machine that worked, and that still works four decades later.

The Atari works in a way like no other computer. In order to produce television graphics and sound on the cheap, Atari designed a custom chip called the Television Interface Adapter (TIA). The TIA makes bizarre demands: instead of preparing a screen's worth of television picture all at once, the programmer had to alter data the TIA transmitted in tandem with the scan-line-by-scan-line movement of the television's electron beam, the process an old cathode ray tube (CRT) television uses to create a visible, two-dimensional picture. Programming the Atari feels more like plowing a field than like painting a picture.

As critics and engineers, Nick and I were interested in the Atari's role in human creativity and culture: how computer hardware influenced game design and aesthetics. You can see the effects of the TIA's line-by-line logic in Atari games: in the rows of targets in Air-Sea Battle or the horizontal bars of horizon in Barnstorming. Methodically, we pointed out these couplings between machine and expression in the popular games of the era, showing how they influenced later designs.

The Atari was made by people in order to entertain other people, and in that sense it's only a machine. But a machine and its components are also something more, something alive, almost, in the way

Kondo endorses. I couldn't help but feel enchanted by the system's parts as much as its output. I found myself wondering, what is it like to be an Atari, or a Television Interface Adapter, or a cathode ray tube television?

Such a question may seem far-fetched or just plain mad. But is it really so strange? Is it really that different from scrutinizing the giant tortoise or the wide body jumbo jet? Or the mall tiles, the lawn, or anything else, for that matter? It has become unfathomable simply to be enamored with things in the world and to embrace such fascination, not to get something back, not to master something, but *just because*. Suspending our suspicion about the use or virtue of a thing—of anything—at least long enough to ask what it is, to treat it as if its existence were reasonable, to acknowledge that it might be incorporable into our experience while also exceeding that experience.

A common scene in my childhood home: my father, sitting on the floor, perched mere inches before our wood-clad, 1970s console television. At that distance, the electric static from its picture tube becomes palpable, a Tesla coil lapping current at your nose like a puppy. It offered the best compromise given his visual limitations. I used to try it too, only to find that the television picture would decompose into the red, green, and blue beams of the color set's individual electron guns, broken down into the pattern of the aperture grill through which they are focused.

Two decades later, rediscovering the Atari with Nick for our book, I realized something. The sensual ether of the television becomes as much an object of concern as the characters and scenes represented upon it, or the creators who fashion them from film or from software, or the viewers and players who watch and manipulate them. Not because it feels fashionable to prefer the fuzzy static of a tiny tube television in the era of high definition, or the blocky nuance of Atari games in the era of Grand Theft Auto, but because the television offers its own intrigue, no matter what it displays.

How deep can one dive into this reality? Not to gain leverage on it, to one-up it, but to commune with it. To nod in its direction, to acknowledge its existence with a tip of a metaphysical hat no one can see. A less anthropocentric version of Epictetus's faux bedside death wish, one that deepens its commitment by refusing the selfish context of genetics or personal loss. To know enough about it that the temptation to ironize it is dissipated by the unseemliness—the violence, even—of suffocating it in proverbial plastic.

It's a procedure conducted by uncovering and understanding its constraints. In the Atari's case: a cathode ray tube fires patterns of electrons at a phosphorescent screen, which glows to create a visible picture. This screen image is not displayed like a photograph, but in individual scan lines created as the electron gun passes from side to side across the screen. Because the Atari has no video memory, its programs must interface between processor and electron gun during every moment of every line of the television display.

For the beginning Atari programmer, it's relatively easy to work scan line by scan line, setting up the next line's worth of sprite and background data and colors in the twenty-two processor cycles it takes for the electron gun to reset its position from one edge of the picture to the other before turning on again. But eventually, the adept creator will need to change something part way through a scan line.

The electron beam continues scanning the phosphor display whether the program does anything or not. The programmer must time microprocessor instructions so that appropriate changes to the picture take place once the first part of a scan line has completed. To do this, you count the number of processor cycles that a set of instructions will take to execute, and match that time up with the amount of horizontal space on the scan line that will have been traced by the electron gun in the meantime. Mistiming by even a single cycle can set the entire display out of whack, rolling horizontally and vertically, like a Dalí painting in video form.

Processor cycle, radio frequency, horizontal blank, electron gun, microchip—to see the richness of the one world, sometimes we must make ourselves blind to another. We must commit to live, even if for a time, within a different magic circle, an unfamiliar playground. We must live there in order to exercise the muscle of defamiliarization. Before we can commune with anything, we must first be able to face it unironically.

Stephen Shore's eye and the resolution of his film plates help elevate his images from half-hearted snapshots to ontographic masterworks, and the same is true of Matthew Weiner's obsessive set and prop design for *Mad Men*. But a similar effect is possible with far more ordinary materials, particularly if we allow them to plug in to the power of play without allowing that power to ratchet up into full-blown ironoia. The result is a rapt and detailed attention we don't seem capable of maintaining under ordinary circumstances. We don't play in order to distract ourselves from the world, but in order to partake in it.

•••

COLONEL STAPP WAS eighty-two years old and suffering from the long-term effects of the many injuries he had sustained to become the fastest man alive when I met him in the summer of 1992, at the International Space Hall of Fame in Alamogordo. I was a camp counselor teaching schoolchildren who fancied that they wanted to touch the vacuum of space, or at least to ponder it for a week. The complex sits at the base of the Sacramento Mountains, a few miles from the stretch of empty desert where Stapp's rocket sleds had traversed the length of a football field in half a second. The museum's Project Gemini–inspired gold-mirrored windows turn away the desert sun, dumping luster onto the courtyard like a tipping mine car. There, Sonic Wind is fastened to the cement yard, immovable, put out to pasture. A relic, a symbol, the Andy Warhol Brillo box of land rocketeering.

As with my father, Stapp's fractured bones had long healed, but detached retinas and permanently broken blood vessels had significantly reduced the clarity of his vision. In exchange, he'd tousled speed and wind, an Icarus with roadrunner wings. The colonel had been aware of the risks—before his final run on Sonic Wind he'd practiced dressing and undressing with the lights out "so if I was blinded I wouldn't be helpless."[2] The ultimate rocket sled experience isn't made of speed and risk, but of the shape and orientation of bone or plastic buttons into the blanket-stitched fabric holes on pressed dress shirts.

Just as sincerity inverts into contempt, so sight overheats and reverses into pride. Blindness ruptures confidence. It recognizes the preposterousness of all things. It reveals that perception within can never fully make sense of things without, and that earnestness is a backward name for selfishness; and likewise, that contempt confuses one's own failings for another thing's authenticity. Blindness is the ink through which ground shifts to become figure. It glints with wonder, wonder at the fact that we can gain any purchase on the world whatsoever, that we can touch even part of its sensuousness, that we can characterize its experience at all, even a little.

And yet we resist. When we talk about "things," we most often mean them as concepts or abstractions in our minds, rather than as, well, *things*—toasters or wind or combine harvesters. As with happiness, things become things of *ours*. "How are things?" you ask a friend. "I dunno," she replies, "things are weird," or "A strange thing happened on the way home," or "The thing I like most about you is that you're so thoughtful to ask!"

There's no doubt that thoughts are things as much as thorns and thoroughfares, but isn't it striking that a word that we'd first associate with matter so quickly becomes one of matters, of affairs, incidents, penchants, circumstances, activities, and notions. Human ones, at that, and ours, individuated. All the ways we live and think in our own heads, in worlds that revolve around us and our little lives, our series of nights on Earth.

And yet those lives are surrounded by stuff of all stripes, all the stuff ironoia demands we resist rather than commune with: Smithfield half hams and cheeseburger-flavored Pringles and Old Spice deodorant, ceramic tabletops and heathered sweaters and quilted metals, cities and meadows and kudzu, evening and chipmunks and silence.

What if the problem is us talking about it so much? Instead of a world of human knowledge or progress or justice or creativity or happiness, let's instead imagine one in which everything possesses as potentially rich and legitimate an existence as anything else. That everyone and everything deserves more than our mere sincerity could possibly give it. That an infinity of potential playgrounds surround us, and not only for our use but for everyone's— and everything's. As the lifelong ironoiac David Foster Wallace puts it, "somehow altering or getting free of my natural, hard-wired default setting which is to be deeply and literally self-centered and to see and interpret everything through this lens of self."[3] Perhaps irony might yet give way to openness and tentativeness, rather than enclosure and finality.

Pier Pasolini helps us take that openness further, beyond our heads and into the world. "The first image of my life," he writes, "is a white, transparent blind, which hangs—without moving, I believe— from a window which looks out on to a somewhat sad and dark lane." But as a creator more acclimated to generous spectatorship, thanks to cinema, ordinary things need not remain subordinate for Pasolini. Instead, just the opposite—they are more likely to be sublime. He continues, "That blind terrifies me and fills me with anguish: not as something threatening and unpleasant but as something cosmic."[4]

Something cosmic. Even window blinds. Even cheeseburger-flavored Pringles. To stave off ironoia, we need not resist the crass material world nor transform it into artisanal affectation. A gentler touch is needed, a more careful physical therapy: to spend time with things, to visit with them, to give them a chance to be exactly what they are. To shut up for a minute and allow the universe to hum without recourse, without appeal to moralism or nihilism, without trying to take it as a petting zoo or a Death Star, without always caging it in Instagrams or problematizing it in think pieces or strip mining it to fuel the ovens of our own contentment.

We could all benefit by being reared by the blind. Living with things requires that we become continuously blind to them, that we exercise the ability to see them fresh, familiar or not, by refusing to allow them to collapse into servants or obstacles. Blindness, fun, play, limit, constraint—all these are synonyms for humility. There between earnestness and cynicism, in the chasm where ironoia throbs, we can also find solace if we are willing to pause long enough to stop scolding things for failing to yield us comfort. This is the pleasure of limits, the fun of play. Not doing what we want, but doing what we can with what is given.

Acknowledgments

I WOULD LIKE to thank Abbey and Tristan and Flannery and Beatrix, and also: Leigh Alexander, Ross Andersen, Doug Armato, Tim Bartlett, Tom Bissell, Tom Boellstorff, Max Brockman, Rob Capps, Heather Chaplin, Michael Clune, Mia Consalvo, Carl DiSalvo, Rob Dubbin, Megan Garber, Jesse James Garrett, J. J. Gould, Graham Harman, Lara Heimert, David Kammerman, Raph Koster, Adrienne LaFrance, Frank Lantz, Alexis Madrigal, Robinson Meyer, Nick Montfort, Evgeny Morozov, Tim Morton, Ben Platt, Kenneth Reinhard, Becca Rosen, Christopher Schaberg, Janelle Schwartz, Doug Sery, Jason Tanz, Clive Thompson, Iain Thomson, McKenzie Wark, Eric Zimmerman, and Eric Zinner.

Notes

ONE: EVERYWHERE, PLAYGROUNDS

1. Fletcher Bascom Dresslar, *Superstition and Education*, 5:1 (Berkeley: University of California Press, 1907), 94.

2. David Foster Wallace, *This Is Water: Some Thoughts, Delivered on a Significant Occasion, about Living a Compassionate Life* (New York: Little, Brown and Company, 2009), 62.

3. Ibid., 84.

4. Actually, I wrote another book about that. Ian Bogost, *Alien Phenomenology, or What It's Like to Be a Thing* (Minneapolis: University of Minnesota Press, 2012).

5. Wallace, *This Is Water*, 6.

6. Ibid., 120.

7. John Wesley Powell, *Physiographic Regions of the United States* (New York: American Book Company, 1895), 65–66.

TWO: IRONOIA, THE MISTRUST OF THINGS

1. Niklas Luhmann, *Art as a Social System*, trans. Eva M. Knodt (Palo Alto: Stanford University Press, 2000), 73.

2. The annual US Bureau of Labor Statistics survey reports on time use are available at www.bls.gov/tus/data.htm. For some of the summaries mentioned here, see Bjarki, "7 Time Consuming Things an Average Joe Spends on in a Lifetime—The Tempo Blog," *The Tempo* (blog), September 5, 2013, http://blog.tempoplugin.com/2013/7-time-consuming -things-an-average-joe-spends-in-a-lifetime/.

3. Pier Paolo Pasolini, *Lutheran Letters*, trans. Stuart Hood (Manchester: Carcanet New Press, 1983), 30.

4. Ibid., 31.

5. Immanuel Kant, *The Critique of Judgment*, trans. Werner S. Pluhar (Indianapolis, IN: Hackett Publishing Company, 1987), 25–29.

6. Roy Christopher, "Ennui Go: Pop Culture's Irony Fatigue," December 25, 2012, http://roychristopher.com/pop-cultures-irony-fatigue.

7. Christy Wampole, "How to Live Without Irony," *New York Times*, November 17, 2012, http://opinionator.blogs.nytimes.com/2012/11/17 /how-to-live-without-irony/.

8. Jess Zimmerman, "Quite Possibly the Best Bike-for-Sale Ad Ever," *Grist*, June 14, 2012, http://grist.org/list/quite-possibly-the-best-bike-for -sale-ad-ever/. Craigslist took the ad down after it was flagged, but its owner, interviewed by *Grist*, insists he really did have a bike to sell.

9. Joe Coscarelli, "A Discussion With Christy Wampole, the New York Times' Earnest Irony-Hater," *New York Magazine*, November 20, 2012, http://nymag.com/daily/intelligencer/2012/11/new-york-times-christy -wampole-hipster-irony-interview.html.

10. Aristotle, *The Nicomachean Ethics* (Indianapolis: Hackett Publishing Company, 1985), 1108a12.

11. David Foster Wallace, *A Supposedly Fun Thing I'll Never Do Again: Essays and Arguments* (Boston: Little, Brown and Company, 1998), 65.

12. Ibid., 46.

13. Ian Crouch, "The Endless T-Shirt Is a Trend," *The New Yorker*, May 22, 2014, http://www.newyorker.com/culture/culture-desk/the-endless -t-shirt-is-a-trend.

14. Timothy Morton, *Realist Magic* (Ann Arbor: University of Michigan/ Open Humanities Press, 2012), 88.

15. Jeffrey Grub, "Phil Fish Tweet Rebuilt in Minecraft," *Venture-Beat*, April 12, 2013, http://venturebeat.com/2013/04/12/phil-fish-tweet-rebuilt-in-minecraft/.

16. Matthew Humphries, "Notch's $70 million Beverly Hills mansion now exists in Minecraft form," *Geek.com*, December 22, 2014, http://www.geek.com/games/notchs-70-million-beverly-hills-mansion-now-exists-in-minecraft-form-1612164/.

17. Aaron Santesso, *A Careful Longing: The Poetics and Problems of Nostalgia* (Newark: University of Delaware Press, 2006), 14.

18. Bogost, *Alien Phenomenology*, 38.

19. Megan Garber, "Reverse Engineering McDonald's: How to Make a Scarily Authentic Filet-o-Fish," *The Atlantic*, March 18, 2013, http://www.theatlantic.com/technology/archive/2013/03/reverse-engineering-mcdonalds-how-to-make-a-scarily-authentic-filet-o-fish/274124/.

20. Jeb Boniakowski, "We Must Build An Enormous McWorld In Times Square, A Xanadu Representing A McDonald's From Every Nation," *TheAwl.com*, January 23, 2013, http://www.theawl.com/2013/01/giant-mcdonalds-times-square.

21. Viktor Shklovsky, *Theory of Prose*, trans. Benjamin Sheer (Victoria, TX: Dalkey Archive Press, 1991), 140. "Enstrange" is translator Sheer's rendering of the Russian word "остранение."

22. Wallace, *A Supposedly Fun Thing I'll Never Do Again*, 65.

23. David Foster Wallace, *Infinite Jest* (New York: Back Bay Books, 1997), 694–695.

THREE: FUN ISN'T PLEASURE, IT'S NOVELTY

1. "♥VIDEOGAMES♥," accessed October 7, 2015, http://harmonyzone.org/Videogames.html.

2. Ludwig Wittgenstein, *Philosophical Investigations*, trans. G. E. M. Anscombe (Oxford: Blackwell Publishers, 1967), 7, 8, 67.

3. Raph Koster, *A Theory of Fun for Game Design*, 2nd ed. (Sebastopol, CA: O'Reilly Media, 2013), 155.

4. Wallace, *A Supposedly Fun Thing I'll Never Do Again*, 25.

5. James Hibberd, "Joss Whedon: The Definitive EW Interview,"

Entertainment Weekly, September 24, 2014, http://www.ew.com/article /2013/09/24/joss-whedon-interview.

6. Mihaly Csikszentmihalyi, *Flow: The Psychology of Optimal Experience* (New York: Harper Perennial, 2008), 3–7.

7. See Marc Augé, *Non-Places: An Introduction to Supermodernity*, 2nd ed., trans. John Howe, (London: Verso Books, 2009).

8. David Foster Wallace, *The Pale King* (New York: Little, Brown and Company, 2012), 136.

9. Martin Heidegger, *Being and Time*, trans. Joan Stambaugh (Albany, NY: State University of New York Press, 1996), 5, 22, 39, 60–75.

10. Marshall McLuhan and Eric McLuhan, *Laws of Media: The New Science* (Toronto: University of Toronto Press, 1988), 15–16, 40–41, 77, 227–228; Marshall McLuhan et al., *Letters of Marshall McLuhan* (New York: Oxford University Press Canada, 1987), 478.

11. Marshall McLuhan, Quentin Fiore, and Jerome Agel, *War and Peace in the Global Village* (New York: Bantam Books, 1968), 175.

12. Jennifer Senior, *All Joy and No Fun: The Paradox of Modern Parenthood* (New York: Ecco Press, 2015), 6.

13. Nancy Darling, "Why Parenting Isn't Fun," *Psychology Today*, July 18, 2010, https://www.psychologytoday.com/blog/thinking-about-kids /201007/why-parenting-isn-t-fun.

14. Johan Huizinga, *Homo Ludens: A Study of the Play-Element in Culture* (Boston: Beacon Press, 1972), 3.

15. Bernard Suits, *The Grasshopper: Games, Life and Utopia* (Peterborough, ON: Broadview Press, 2005), 157.

16. Hubert L. Dreyfus and Sean Dorrance Kelly, *All Things Shining: Reading the Western Classics to Find Meaning in a Secular Age* (New York: Free Press, 2011), chapter 2 *passim*.

17. Wallace, *The Pale King*, 436.

FOUR: PLAY IS IN THINGS, NOT IN YOU

1. Jacques Derrida, *Writing and Difference*, trans. Alan Bass (Chicago: University of Chicago Press, 1980), 278.

2. Brian Sutton-Smith, *The Ambiguity of Play* (Cambridge, MA: Harvard University Press, 2001), 182.

3. Katie Tekinbaş and Eric Zimmerman, *Rules of Play: Game Design Fundamentals* (Cambridge, MA: MIT Press, 2003), 304.

4. Peter Gray, "The Play Deficit," *Aeon Magazine*, September 18, 2013, http://aeon.co/magazine/culture/children-today-are-suffering-a-severe-deficit-of-play/.

5. David Derbyshire, "How Children Lost the Right to Roam in Four Generations," *Daily Mail*, June 15, 2007, http://www.dailymail.co.uk/news/article-462091/How-children-lost-right-roam-generations.html.

6. "Parents in Trouble Again for Letting Kids Walk Alone," *USA Today*, April 13, 2015, http://www.usatoday.com/story/news/nation/2015/04/13/parents-investigated-letting-children-walk-alone/25700823/.

7. Stuart Brown, "Play Is More than Just Fun" (lecture, TED Talk, Serious Play Studios, Pasadena, CA, 2008), http://www.ted.com/talks/stuart_brown_says_play_is_more_than_fun_it_s_vital.

8. Sutton-Smith, *The Ambiguity of Play*, 198. The actual quote is less quotable: "What is adaptive about play, therefore, may be not only the skills that are a part of it but also the willful belief in acting out one's own capacity for the future. The opposite of play, in these terms, is not a present reality or work, it is vacillation, or worse, it is depression." How the more succinct version arose is a mystery, but the original betrays the simplicity and definitiveness common in its citation.

9. Huizinga, *Homo Ludens*, 10–11.

10. Miguel Sicart, "How the 21st Century Will Be Defined by Games," *Wired.co.UK*, February 4, 2015, http://www.wired.co.uk/magazine/archive/2015/04/ideas-bank/ludic-century.

11. Gray, "The Play Deficit."

12. Marshall Sella, "Against Irony," *New York Times Magazine*, September 5, 1999, https://partners.nytimes.com/library/magazine/home/19990905mag-sincere-culture.html.

13. Miguel Sicart, *Play Matters* (Cambridge, MA: MIT Press, 2014), 36.

14. Ibid., 38.

15. Mary Flanagan, *Critical Play: Radical Game Design* (Cambridge, MA: MIT Press, 2013), 14, 32–33.

16. Huizinga, *Homo Ludens*, 10.

17. Ruth Graham, "How the American Playground Was Born in

Boston," *The Boston Globe*, March 28, 2014, https://www.bostonglobe.com /ideas/2014/03/28/how-american-playground-was-born-boston/5i2Xr MCjCkuu5521uxleEL/story.html.

18. Tenkinbaş and Zimmerman, 95.

19. Sicart, *Play Matters*, 40.

20. Koster, *A Theory of Fun for Game Design*, 39–40.

21. Suits, *The Grasshopper*, 54–55.

FIVE: FROM RESTRAINT TO CONSTRAINT

1. Barry Schwartz, *The Paradox of Choice: Why More Is Less* (New York: HarperCollins Publishers, 2005), 17.

2. Jessica Salter, "The Man Who Lives Without Money," *The Telegraph*, August 18, 2010, http://www.telegraph.co.uk/news/earth/greenerliving /7951968/The-man-who-lives-without-money.html.

3. James Hamblin, "Living Simply in a Dumpster," *The Atlantic*, September 11, 2014, http://www.theatlantic.com/features/archive/2014/09/the -simple-life-in-a-dumpster/379947/.

4. Katy McLaughlin, "How We Began Enjoying More by Indulging Less," *Wall Street Journal*, November 4, 2012, http://www.wsj.com/articles /SB10001424052970204840504578086630967835850.

5. "30 Surprising Facts About How We Actually Spend Our Time," *Distractify*, January 7, 2015, https://www.distractify.com/astounding-facts -about-how-we-actually-spend-our-time-1197818577.html.

6. Sherry Turkle, *Reclaiming Conversation: The Power of Talk in a Digital Age* (New York: Penguin, 2015), 3.

7. "How 'The Phone Stack' Is Civilizing Dinners out with Friends," *The Week*, January 10, 2012, http://theweek.com/articles/479030/how -phone-stack-civilizing-dinners-friends.

8. Marie Kondo, *The Life-Changing Magic of Tidying Up: The Japanese Art of Decluttering and Organizing* (Berkeley, CA: Ten Speed Press, 2014), 1–5.

9. Antoine de Saint-Exupéry, *Wind, Sand and Stars*, trans. Lewis Galan-tière (New York: Mariner Books, 2010), 42.

10. Kondo, *The Life-Changing Magic of Tidying Up*, 41.

11. Ibid.

12. Ibid., 81.

13. Lauren Sherman, "The KonMari Method Is a Boon to Second-hand Stores and E-tailers," *Fashionista*, April 6, 2015, http://fashionista.com/2015/04/konmari-method-goodwill.

14. George Carlin, *3 x Carlin: An Orgy of George* (New York: Hachette Books, 2015).

15. Timothy Morton, *Hyperobjects: Philosophy and Ecology after the End of the World* (Minneapolis: University of Minnesota Press, 2013), 115.

16. Carlye Wisel, "What If All My Stuff Sparks Joy? The Problem with Marie Kondo," *Racked*, May 18, 2015, http://www.racked.com/2015/5/18/8607043/marie-kondo-konmari-cleaning-tidying-advice.

17. See https://freedom.to.

18. Megan Rose Dickey, "To Protect His Livelihood, This Guy Locks His Smartphone in a Safe Every Day," *Business Insider*, March 15, 2013, http://www.businessinsider.com/guy-locks-router-and-smartphone-in-safe-2013-3.

19. Kriston Capps, "When Micro-Housing Misses the Point," *CityLab*, June 24, 2014, http://www.citylab.com/housing/2014/06/when-micro-housing-misses-the-point/373104/.

20. Penelope Green, "One Shed Fits All: A Modernist Dogtrot Reborn," *New York Times*, October 4, 2012, http://www.nytimes.com/2012/10/04/greathomesanddestinations/one-shed-fits-all-a-modernist-dogtrot-reborn.html?_r=0.

21. Richard A. Easterlin, "Does Economic Growth Improve the Human Lot? Some Empirical Evidence," *Nations and Households in Economic Growth*, 1974, 89–125.

22. Daniel Kahneman and Angus Deaton, "High Income Improves Evaluation of Life but Not Emotional Well-Being," *Proceedings of the National Academy of Sciences* 107, no. 38 (September 7, 2010): 16489–16493.

23. Leaf Van Boven and Thomas Gilovich, "To Do or to Have? That Is the Question," *Journal of Personality and Social Psychology* 85, no. 6 (2003): 1193–1202.

24. Jay Cassano, "The Science of Why You Should Spend Your Money on Experiences, Not Things," *FastCo.Exist.com*, March 30, 2015, http://www.fastcoexist.com/3043858/world-changing-ideas/the-science-of-why-you-should-spend-your-money-on-experiences-not-thing.

25. Darwin A. Guevarra and Ryan T. Howell, "To Have in Order to Do:

Exploring the Effects of Consuming Experiential Products on Well-Being," *Journal of Consumer Psychology* 25, no. 1 (January 2015): 28–41.

26. Jerome J. McGann, *Black Riders: The Visible Language of Modernism* (Princeton: Princeton University Press, 1993), 114.

27. William Morris, "Textiles," *Journal of the Society of Arts* 36 (November 18, 1887–November 16, 1888), 1133.

28. Igor Stravinsky, *Poetics of Music in the Form of Six Lessons* (Cambridge, MA: Harvard University Press, 1970), 63.

29. Alfred North Whitehead, *Religion in the Making* (Cambridge: Cambridge University Press, 1926), 77–79; Alfred North Whitehead, *Process and Reality* (New York: The Free Press, 2010), 21, 31–32.

30. Michael Halewood, "There Is No Meaning to 'Creativity' Apart from Its 'Creatures'" (paper presented at a Unit of Play Seminar, Goldsmiths, University of London, January 10, 2013).

31. Whitehead, *Process and Reality*, 31.

32. Stravinsky, *Poetics of Music in the Form of Six Lessons*, 65.

33. Maria Popova, "Charles Eames on What Is Design," *The Atlantic*, October 5, 2011, http://www.theatlantic.com/entertainment/archive/2011/10/charles-eames-on-what-is-design/246034/.

SIX: THE PLEASURE OF LIMITS

1. Homer, *The Odyssey*, trans. Robert Fagles (New York: Penguin, 1999), book I, lines 1–2.

2. William Shakespeare, *Shakespeare's Sonnets* (Folger Shakespeare Library), eds. Barbara A. Mowat and Paul Werstine (New York: Simon & Schuster, 2004), 37.

3. Ben Dobbin, "Toy Hall of Fame Points to New Addition: The Stick," *USA Today*, November 7, 2008, http://usatoday30.usatoday.com/news/2008–11–07–3791182869_x.htm.

4. Donald A. Norman, *The Design of Everyday Things* (New York: Basic Books, 2002), 10–13, 60–62, 82–84, 87–88.

5. Ibid., vii–viii, 2–5.

6. James J. Gibson, *The Ecological Approach to Visual Perception* (New York: Psychology Press, 1979), 15.

7. E.g., Kofi Kissi Dompere, *The Theory of the Knowledge Square: The*

Fuzzy Rational Foundations of the Knowledge-Production Systems (New York: Springer, 2012), 36.

8. Adam Gopnik, "Why Teach English?," *The New Yorker*, August 27, 2013, http://www.newyorker.com/books/page-turner/why-teach-english.

9. Pew Research Center, "E-Reading Rises as Device Ownership Jumps" (January 2014), http://pewinternet.org/reports/2014/E-Reading-Update .aspx.

10. Scholastic, "Kids & Family Reading Report," 4th Edition, http:// mediaroom.scholastic.com/files/kfrr2013-noappendix.pdf.

11. Miles Kimball and Noah Smith, "The Myth of 'I'm Bad at Math,'" *The Atlantic*, October 28, 2013, http://www.theatlantic.com/education /archive/2013/10/the-myth-of-im-bad-at-math/280914/.

12. Norman, *The Design of Everyday Things*, 188.

13. Alice Maggin, "Dr. Seuss Classic Hits Milestone," *ABC News*, August 13, 2010, http://abcnews.go.com/WN/dr-seuss-green-eggs-ham-50th -anniversary-beloved/story?id=11384227.

14. Philip Nel, *Dr. Seuss: American Icon* (New York: Continuum, 2004), 29–30.

15. Lynn Neary, "Fifty Years of 'The Cat in the Hat,'" *NPR.com*, March 1, 2007, http://www.npr.org/2007/03/01/7651308/fifty-years-of -the-cat-in-the-hat.

16. Lydia Maria Francis Child, *The American Frugal Housewife: Dedicated to Those Who Are Not Ashamed of Economy*, 12th ed. (Boston: Applewood Books, 1989), 82.

17. Ibid., 89.

18. Brian X. Chen, "In Busy Silicon Valley, Protein Powder Is in Demand,"*New York Times*, May 26, 2015, http://www.nytimes.com/2015 /05/25/technology/in-busy-silicon-valley-protein-powder-is-in-demand .html.

19. "Heads Up!" *Chopped* 15:1, March 31, 2013.

20. Ibid.

21. Huizinga, *Homo Ludens*, 11. The same phrase repeats with greater urgency later: "the cheat or the spoil-sport shatters civilization itself" (211).

22. Joel Cunningham, "8 Best-Sellers Started During National Novel

Writing Month," *B&NReads* (blog), November 1, 2013, http://www
.barnesandnoble.com/blog/8-best-sellers-started-during-national-novel
-writing-month/.

23. Malcolm Gladwell, *Outliers: The Story of Success* (New York: Little,
Brown & Company, 2011), chapter 2 *passim*.

24. Georges Perec, "Le Grand Palindrome," *La littérature potentielle:
créations, re-créations, recreations* (Paris: Gallimard, 1973). You can read it
online at http://homepage.urbanet.ch/cruci.com/lexique/palindrome
.htm.

25. Martin Gardner, *Penrose Tiles to Trapdoor Ciphers: And the Return of
Dr. Matrix* (Washington, DC: The Mathematical Association of America,
1997), 83.

26. Nick Montfort and William Gillespie, *2002: A Palindrome Story in
2002 Words* (Urbana, IL: Spineless Books, 2002), 1.

27. Georges Perec and Georges Pérec, *La Disparition: Roman* (Paris:
Denoel, 1969).

28. Georges Perec, *A Void*, trans. Gilbert Adair (London: The Harvill
Press, 1994).

29. The demo page is at http://www.pouet.net/prod.php?which
=52938, but you might prefer to watch a YouTube video, https://www
.youtube.com/watch?v=WGf4kU0pIZk.

30. http://www.ioccc.org/1990/westley.c.

31. Michael Mateas and Nick Montfort, "A Box, Darkly: Obfuscation,
Weird Languages, and Code Aesthetics" (paper presented at Digital Arts
and Culture: Digital Experience: Design, Aesthetics, Practice), Copenha-
gen, Denmark, 2005, https://users.soe.ucsc.edu/~michaelm/publications
/mateas2-dac2005.pdf, 5–6.

32. The author's partial explanation is here, http://www.ioccc.org
/1990/westley.hint.

33. http://www.dangermouse.net/esoteric.

34. David Sarno, "Twitter Creator Jack Dorsey Illuminates the Site's
Founding Document. Part I," *LA Times*, February 18, 2009, http://
latimesblogs.latimes.com/technology/2009/02/twitter-creator.html.

35. Heather Kelly, "OMG, the Text Message Turns 20. But has SMS
Peaked?" *CNN*, December 3, 2012, http://www.cnn.com/2012/12/03/tech
/mobile/sms-text-message-20/.

36. Derek Thompson, "The Unbearable Lightness of Tweeting," *The Atlantic*, February 16, 2015, http://www.theatlantic.com/business/archive/2015/02/the-unbearable-lightness-of-tweeting/385484/.

37. Kurt Wagner and Jason Del Rey, "Twitter Plans to Go Beyond Its 140-Character Limit," *Re/code*, September 29, 2015, http://recode.net/2015/09/29/twitter-plans-to-go-beyond-its-140-character-limit/; Yoree Koh, "Twitter to Expand Tweet's 140-Character Limit to 10,000," *Wall Street Journal*, January 5, 2016, http://blogs.wsj.com/digits/2016/01/05/twitter-to-expand-tweets-140-character-limit-to-10000/. In July 2015, Twitter did remove the limitation from direct messages, but mostly to facilitate the corporate customer service. In a March 2016 appearance on NBC's *Today Show* marking Twitter's tenth anniversary, CEO Jack Dorsey told Today host Matt Lauer that his company would "absolutely not" remove the 140 character restriction: Eun Kyung Kim, "Twitter CEO Jack Dorsey on whether platform censors users: 'Absolutely not'" *TODAY Money*, March 18, 2016, http://www.today.com/money/twitter-ceo-jack-dorsey-whether-platform-censors-users-absolutely-not-t81181.

38. Natasha Dow Schüll, *Addiction by Design: Machine Gambling in Las Vegas* (Princeton, NJ: Princeton University Press, 2014), 2.

39. Luke Stark and Kate Crawford, "The Conservatism of Emoji," *The New Inquiry*, August 20, 2014, http://thenewinquiry.com/essays/the-conservatism-of-emoji/.

SEVEN: THE OPPOSITE OF HAPPINESS

1. Jason Tanz, "The Curse of Cow Clicker: How a Cheeky Satire Became a Videogame Hit," *Wired* 20.01 (January 2012): 118, http://archive.wired.com/magazine/2011/12/ff_cowclicker/all/.

2. Ian Bogost, "Shit Crayons," accessed March 3, 2011, http://bogost.com/writing/shit_crayons/.

3. Tanz, "The Curse of Cow Clicker," 118.

4. Ralph Waldo Emerson, *Works* (London: Routledge, 1883), 564.

5. Barbara Ehrenreich, *Smile or Die: How Positive Thinking Fooled America and the World* (London: Granta Books, 2010), 8.

6. http://www.happify.com/hd/science-of-happiness-infographic/.

7. Sonja Lyubomirsky, *The How of Happiness: A New Approach to Getting the Life You Want* (New York: Penguin, 2008), 20.

8. Nicholas J. L. Brown, Harris L. Friedman, and Alan D. Sokal, "The Complex Dynamics of Wishful Thinking: The Critical Positivity Ratio,"*American Psychologist* 68, no. 9 (2013): 801–813.

9. Laura Clark, "How Halitosis Became a Medical Condition With a 'Cure,'"*Smithsonian*, January 29, 2015, http://www.smithsonianmag.com /smart-news/marketing-campaign-invented-halitosis-180954082/?no-ist.

10. Jackson Lears, "Get Happy!!," *The Nation*, November 6, 2013, http:// www.thenation.com/article/get-happy-2/.

11. Ada S. Jaarsma, "An Existential Look at *Mad Men:* Don Draper, Advertising, and the Promise of Happiness," in *Mad Men and Philosophy: Nothing Is as It Seems*, eds. Rod Carveth, William Irwin, and James South (Hoboken, NJ: Wiley, 2010), 99–102.

12. Kondo, *The Life-Changing Magic of Tidying Up*, 203.

13. Oliver Burkeman, *The Antidote: Happiness for People Who Can't Stand Positive Thinking* (London: Faber And Faber, 2013), 33.

14. Epictetus, *Discourses*, 3.24.88.

15. Kondo, *The Life-Changing Magic of Tidying Up*, 40.

16. John Maeda, *The Laws of Simplicity* (Cambridge, MA: MIT Press, 2006), 69.

17. Bruno Latour, *We Have Never Been Modern*, 3rd ed., trans. Catherine Porter (Cambridge, MA: Prentice Hall/Harvester Wheatsheaf, 1993), 30.

18. Ibid., 144.

19. Alan Watts, *The Wisdom of Insecurity* (New York: Pantheon, 1952), 81.

20. Burkeman, *The Antidote*, 148.

21. Ibid., 208–209.

CONCLUSION: LIVING WITH THINGS

1. Nick Montfort and Ian Bogost, *Racing the Beam: The Atari Computer System* (Cambridge, MA: MIT Press, 2009).

2. Associated Press, "'Rocket Sled' Rider Of '54 Gratified By Results," *New York Times*, December 26, 1984, http://www.nytimes.com/1984/12/26 /us/rocket-sled-rider-of-54-gratified-by-results.html.

3. Wallace, *This Is Water*, 2.

4. Pasolini, *Lutheran Letters*, 28.

Index

GREGORY MILLER

IAN BOGOST is the Ivan Allen College Distin-
guished Chair in Media Studies and a professor
of interactive computing at the Georgia Institute
of Technology, a founding partner at Persuasive
Games, and a contributing editor at *The Atlantic*.
Bogost lives in Atlanta, Georgia.